SpringerBriefs in Electrical and Computer Engineering

For further volumes:
http://www.springer.com/series/10059

Stefano Bellucci · Bhupendra Nath Tiwari
Neeraj Gupta

Geometrical Methods for Power Network Analysis

 Springer

Stefano Bellucci
Laboratori Nazionali di Frascati
Istituto Nazionale di Fisica Nucleare
Frascati, Rome
Italy

Neeraj Gupta
Department of Electrical Engineering
Indian Institute of Technology Kanpur
Kanpur, UP
India

Bhupendra Nath Tiwari
Laboratori Nazionali di Frascati
Istituto Nazionale di Fisica Nucleare
Frascati, Rome
Italy

ISSN 2191-8112   ISSN 2191-8120   (electronic)
ISBN 978-3-642-33343-9   ISBN 978-3-642-33344-6   (eBook)
DOI 10.1007/978-3-642-33344-6
Springer Heidelberg New York Dordrecht London

Library of Congress Control Number: 2012948426

Springer is part of Springer Science+Business Media (www.springer.com)

To our parents and teachers

Preface

This book presents an intrinsic geometric model for power system planning and operation. This problem is generally large scale and nonlinear. We have thus developed an intrinsic geometric model for network reliability and voltage stability, and applied it specifically to the IEEE 5 bus system. The robustness of the proposed model is illustrated by introducing variations of the network parameters. Exact analytical results demonstrate the accuracy as well as the efficiency of the proposed solution technique.

Moreover, from the perspective of intrinsic state-space geometry, we explore power system instability problems introduced by the competitive market mechanism, system failures, and contingency effects, i.e., unforeseen and unexpected events involving system devices, in the face of the widespread deregulation of the global power industry. From the perspective of electrical engineering, we examine the state-space formulation pertaining to:

- voltage regulation phase shift corrections,
- voltage instability of the maximum deliverable power for a given load in the steady-state sinusoidal regime, and
- stabilization of a large-scale voltage instability under the power law characterization.

Finally, regarding the formulation of the intrinsic geometry, we offer a detailed account of complex power optimization in both the steady and nonsteady state regimes. The book can be summarized as follows.

In Chap. 1, we explain the motivation for power network flow analysis and introduction of the geometrical method. The goal of this research is to advance the state of the art in power system reliability and voltage stability. Indeed, power system stability is a potentially attractive target in the emerging deregulated power market, not only to maintain efficient operation of power systems while meeting the demand, but also to optimize the economics of the power system. In addition, deregulation of the electricity market has introduced power flow uncertainty, and hence also fluctuations. This means that power system planners and operators are required to run electricity network components close to their physical capacities, under a given equilibrium operation point. Our scheme maintains system performance with almost stable system voltage profile and transmission line efficiency by improving impedance angle and steady-state controls and by avoiding blackouts. In this respect, planning aimed at providing a stable power supply has recently been in wide demand in power system technology. From the perspective of transmission theory, we report new developments involving power flow properties as a function of the power factors of a finite parameter network configuration.

In Chap. 2, we show how to formulate the intrinsic geometric characterization of admissible component(s) and discuss the associated theoretical motivations from the perspective of circuit planning, and the validity of the proposed model.

In Chap. 3, we provide an intrinsic geometric characterization by incorporating the proposed methodology. We begin with a brief introduction to the network fluctuation problem, recalling the underlying motivation for the intrinsic geometric analysis, and setting up the notation for the computations in subsequent chapters. Implementing the notation for a given network local equilibrium, we can fix a set of optimal values of the network parameters, e.g., L, C, and R, and the corresponding phases of the chosen power network. The logic simply follows from the fact that the sum of the three angles of a triangle is constant. To illustrate these intrinsic geometric considerations, we derive exact formulas for the case of the two- and three-parameter statistical configurations. We also outline the above notion for a network with finitely many parameters.

Chapter 4 discusses current research trends and knowledge of the intrinsic geometric model in an accessible manner for application to the solution of power system planning and operation. This problem is large scale and nonlinear, in general. In order to test network reliability, we have developed an intrinsic geometric model for network reliability and examined it for the IEEE 5 bus system at the forefront of research on the physical aspects of local and global flow properties. In fact, the robustness of the proposed model is illustrated by introducing variations of the network parameters. The exact analytical results here offer a compact and up-to-date discussion of the accuracy and efficiency of the statistical technique.

Chapter 5 extends the aforementioned techniques of intrinsic geometry and examines the problem of voltage stability in network theory. In order to carry out this investigation, we formulate the problem of voltage stability by following the previously described mathematical specifications of the three-parameter model. Hence, in order to test for voltage stability, we describe this innovation for single component LCR networks and show that the voltage stability achieved using the results proven by the proposed work lies within the desired accuracy limits. From the outset of the present investigation, we offer specific remarks and outlook for further research prospects.

From the perspective of network theory, Chap. 6 illustrates how the parametric intrinsic geometric description exhibits an exact set of pair correction functions and global correlation volume, with and without the inclusion of the imaginary power flow. The Gaussian fluctuations about the equilibrium basis for the phases of a power network system generate a well-defined, nondegenerate, curved regular intrinsic Riemannian surface for the purely real and the purely imaginary power flows and their linear combinations. An explicit computation demonstrates that the underlying real and imaginary power correlations involve ordinary summations of the power factors, with and without their joint effects. The novel aspects of intrinsic geometry allow one to propose stable designs for power systems.

In Chap. 7, we examine the phase shift correction under the hypotheses of fluctuation theory. In particular, from the state-space perspective for the voltage instability of a connected power system, we illustrate how, for 2 bus systems, the voltage of the relevant buses, i.e., bus 1 and bus 2, varies in the same way as the transmitted power. In the sequel, we explore the state-space formulation pertaining to voltage regulation and phase shift correction. In the state-space formulation, some assumptions have been made, i.e., the shunt admittance has been neglected, no reactive support on the load bus is allowed, the generator terminal voltage phasor is assumed to coincide with the rotor position, and the load is defined by the real and reactive power demand.

In Chap. 8, we analyze the voltage instability pertaining to the maximum deliverable power and the complex power optimization problem for a given load. We thus illustrate the role of state-space geometry in complex power flow optimization. In the fast-growing and competitive power market, optimization is essential in order to define the loadability limit of the power network. Here, one must consider not only the real power, but also the reactive support.

Chapter 9 uses the intrinsic state-space geometry to explore power system instability problems introduced by the competitive market mechanism, system faults, and contingency of instruments in the face of the widespread deregulation of the global power industry. From the standpoint of electrical engineering, we examine the state-space formulation using the notions from (1) the voltage regulation phase shift correction and (2) the voltage instability of the maximum deliverable power for a given load in the steady-state sinusoidal regime to discuss the stabilization of the large-scale voltage instability under the power law characterization. In the proposed formulation of this problem, the same assumptions are made as in Chap. 7. Finally, regarding the intrinsic geometry formulation, we offer a detailed account of complex power optimization of large-scale voltage instability in both the steady and non-steady state regimes.

Chapter 10 brings together the conclusions from the intrinsic geometric approach for the power network stability problem and discusses the outlook for future research in the subject. The main highlight of the present intrinsic geometric model is that it can be applied to analyze random variations of circuit parameters. It can be used for planning and operation of networks, so a variational analysis can be described from the perspective of operation and maintenance. For a unified analysis of electrical networks, the proposed geometric model offers simultaneous consideration of the following three scenarios: (1) lossless power networks, (2) networks with power loss, and (3) voltage collapses occurring in the network.

The present model is technically sound and could be directly implemented by the technological application of intrinsic Weinhold geometry. Furthermore, such models are robust and are intrinsically parameterized in terms of the reactance, resistance, and capacitance of the considered network. Since we offer a nonlinear improvement of the stability and reliability of the power system, the present investigation eliminates the disadvantages inherent in standard regression techniques. Moreover, the voltage instability results from the fact that the operating point lies beyond the maximum deliverable power. Beyond this limit, the maximum deliverable power leads to a varying load. Thus, a more realistic outage regarding the deliverable power flow is obtained when one takes into account (as in the present nonlinear analysis) fluctuations in the voltage, occurring along with fluctuations in the input real and imaginary power. Finally, in the previously mentioned hypothesis of complex power optimization (see for instance Chap. 8), there exists a set of computational disadvantages, viz., integration with the existing methodologies, lack of economic analysis, and lack of sensitivity analysis, together with possible combined effects. By considering the state-space geometry, several of these problems are nonlinearly improved. Using the results of the present model, it should be possible to produce an engineering demonstrator application. In short, this book gives an accurate load-shedding strategy with respect to network reliability and system instability.

As a bridge between advanced graduate study and the forefront of engineering research, we offer here a detailed examination of intrinsic geometry, network analysis, and statistical configurations aimed at postgraduate students and non-specialist researchers in physics and related applied areas.

Regarding the present research, BNT would like to thank Prof. V. Ravishankar, Prof. P. Jain, Prof. U. B. Tewari, Prof. M. K. Harbola, Prof. R. K. Thareja, and Prof. S. G. Dhande for their encouragement and support while this research was underway at the Indian Institute of Technology in Kanpur, India. The project was initiated while BNT was supported by the Council of Scientific and Industrial Research in New Delhi, India, under the doctoral research fellowship CSIR-SRF-9/92(343)/2004-EMR-I, and completed during a postdoctoral research fellowship at the INFN-Laboratori Nazionali di Frascati in Rome, Italy. NG would like to thank Prof. Prem K. Kalra and Prof. R. Shekhar for the basic motivation, as well as the continued guidance and support for the completion of this work. Finally, we would like to thank our parents and family members for their constant support since the commencement of this research.

Frascati, Italy, July 2012   S. Bellucci
                                                                        B. N. Tiwari
Kanpur, India, July 2012   N. Gupta

Contents

Chapter 1
Introduction

The goal of this research review is to describe advances in the state of the art with regard to power system reliability [1] and voltage (in)stability [2, 3]. We consider a given network configuration in the sense of statistical mechanics and examine the domains of (in)stability from the standpoint of intrinsic geometry. We introduce the geometric theory of statistical stability for power networks. In this respect, it is well known that, for effective power system planning, the appropriate reactive compensation [4] is essential with a suitable set of network parameters (resistance R and reactance X) and associated planning issues [5, 6]. With a given network as the statistical system, such planning helps to reduce the apparent power by cutting reactive power losses in the network.

In particular, in the varying reactive load scenario [7, 8], where stability of the voltage profile is necessary, the addition of the reactive components [9] of a desired bus or line makes a significant difference from the point of view of network design [10–13]. In consequence, it is essential to choose a finite value of the capacitance and reactance to keep the voltage within the stability limit. The value of such parameters leads to the design of an interesting set of power flow protection devices, e.g., tap changer, controlled switched compensator [14, 15]. A compensation strategy is used to improve the voltage profile and power factor correction in AC transmission lines. Note also that the concept of reliability and voltage stability is related to power quality issues [4, 16]. Most power quality issues are resolved by appropriate reactive power [17], which is itself controlled by a capacitor bank [18, 19]. Such a scheme maintains system performance, with an almost stable system voltage profile and efficiency of the transmission line, by improving impedance angle and steady-state controls and avoiding blackouts.

From the perspective of power factor correction and a highly disturbed bus (gradual or abrupt change of reactive load), an appropriate choice of stabilizing compensation is essential [20–22]. With increasing complexity of the power system, the available linear methods yield a slow convergence (or even non-convergence). This follows from the fact that linear analysis involves a number of unrealistic assumptions, due to the uncertain behavior of the load and the contingency of the transmission lines. Thus, it is necessary to go beyond the linear analysis approximation.

S. Bellucci et al., *Geometrical Methods for Power Network Analysis,*
SpringerBriefs in Electrical and Computer Engineering,
DOI: 10.1007/978-3-642-33344-6_1, © The Author(s) 2013

In this respect, our intrinsic geometric technique offers a self-consistent non-linear optimal solution and provides a way round these disadvantages. In addition, there are myriad methodologies and mathematical models for the selection of network parameters and compensation in the area of power system planning and operations, e.g., optimization techniques, genetic algorithms (GA), trial and error methodologies, fuzzy integrated with dynamic programming, artificial neural networks (ANN), and heuristic analysis [23–26], and these have been used in most of the existing software [27, 28]. It is worth emphasizing that precise estimation of transmission line parameters and associated compensation could more efficiently hinder the occurrence of blackouts in power systems.

Using the linear approximation, previous solutions led to results that later proved to be inadequate. This demonstrates the need to incorporate the underlying non-linear effects of parametric fluctuations. In this respect, our intrinsic geometric approach provides ample room to improve network performance through the stable voltage profile of the power system. The previously exploited methodologies yield a set of approximate solutions, which are mostly accomplished iteratively. Such an implementation is realized generically by linearizing the corresponding flow equations in a limited domain. Using the above class of methodologies, the designed compensations are not efficient enough to meet the needs of modern society, due to their slow convergence.

From the very outset of our geometric approach, the voltage stability of the network and safety modes are an immediate consequence of the admissible network parameterization. With the help of correlation techniques [29–32] and identifying the critical point of an arbitrary network, the set of appropriate parameters and components can be identified for any finite component network.

The present geometric model is an optimized control analysis with the ultimate aim of determining all network parameters. Indeed it can provide strategic planning criteria for the most effective use of a network and the issue of network reliability. From the very outset of our geometric approach, the voltage stability of the network and safety modes are an immediate consequence of the admissible network parameterization. Using correlation techniques [29–32] and determining the critical point(s) of an arbitrary network by extremization techniques, the set of appropriate parameters and components can be identified for any finite component network.

Interestingly, the global quantities of our model provide the required set of safety alerts for both the owner and the regulator of the network. For a given total (complex) power, the proposed model is advocated as the next step in power network innovation. The overall methodology is implemented on the IEEE 5-bus system and the results are demonstrated in terms of the key features of the proposed geometric model. The voltage levels of all the buses are assumed equal. This condition holds by the concept of reverse engineering, where the results obtained show that planning could be based on the calculated parameters, which keep the power system at a fixed voltage profile about which network fluctuations are analyzed.

Chapters 2 and 3 detail the techniques summarized in the present chapter. In order to carry out this investigation in a practical setup, we first give the mathematical formulation of the problem and then describe the details of the innovation, showing that it works in general. We then evaluate the results in order to test network reliability and voltage stability by proving that the present proposal can work to within a chosen accuracy. We make specific remarks regarding the present investigation, then draw conclusions and sketch the outlook for future investigations.

On the other hand, planning methods for stable power supply in transmission and distribution systems have been widely applied for a decade now [20]. In the transmission theory, this shows that the power flow varies as a function of the power factor of the network. In order to regulate the voltage of the network, the power factor ϕ is thus related to the impedance angle

$$a \geq \frac{\pi}{2} - \phi. \tag{1.1}$$

Interestingly, a can be determined by tuning network parameters, i.e., resistance r, inductance L, and capacitance C [17, 19]. An optimal choice of a improves the efficiency of power networks, and thus guides toward an appropriate design of network parameters. For a given transmission line or lines, we provide suitable network planning for operation with a minimum reactive power requirement. The present research using intrinsic geometry determines the required unit of power flow under fluctuations of the power factors at the very outset.

Up to now, most network planning and design has been based on the power flow equations [18, 38], so network characterizations are linearly afforded by chronological data analysis, heuristic methods, parametric estimates, and optimization techniques [1, 5]. In this investigation, this motivates us to define non-linear criteria for electrical networks. We thus take into account the fact that the set of voltages at all buses reaches an equilibrium configuration, whence extremization of the power flow in the network can be determined in terms of a. The selection of network parameters determined by optimization techniques could be unreliable and cause bottle-necking. This shows an urgent need for compensator(s). For safe operation and optimal power flow, our analysis provides stability criteria for the non-linearly reliable behavior of the electrical network. In this respect, our method provides a high level compensation strategy to reduce the fluctuation effects of the network parameters, viz., r, L, and C.

To be specific, we focus on power networks and determine the required set of voltage stability criteria, viz., the selection of power factors, network planning, and compensation strategies. Moreover, intrinsic geometry offers an effective route to power system characterization, power factor corrections, and voltage regulation. It is worth mentioning that the voltage level at the buses could be tuned in equilibrium. In the reverse engineering, the proposed solution keeps the power system in a stable voltage range under the fluctuations of the network configuration. In this way, we provide an efficient power system characterization, which is non-linearly stable over the fluctuations of the network phases.

In this context, intrinsic geometry has been important in investigations of fluctuations about an equilibrium configuration. Besides several general notions analyzed in condensed matter physics [33–37], we consider specific analogous electrical networks. As mentioned above, we analyze the parametric pair correlation functions and their correlation relations about the equilibrium configuration. We find that intrinsic geometric considerations entail an intriguing feature of the underlying fluctuations, which are defined in terms of the network parameters.

Given a definite covariant intrinsic geometric description of the network configuration, we expose (i) conditions for stability, (ii) properties of the parametric correlation functions, and (iii) scaling relations in terms of the parameters of the network. In this analysis, we list the complete set of non-trivial parametric correlation functions for the electrical network. Thus, intrinsic geometry plays an important role in the study of power networks and their stable design applications.

In the context of power flow equations [18, 38], the usefulness of the present investigation may seem somewhat limited. However, we show that our approach could in principle be applied generically to all electrical networks, in order to achieve the best possible understanding of the phenomenon and importance of controlled power flow and network analysis. These intrinsic geometric considerations can provide strategic planning criteria for the effective use of power systems and voltage stability. We show that this notion follows from the standard laws of electrical circuits [33]. For an additional component, the criteria of voltage stability via the notion of intrinsic geometric power flow can subsequently be used for optimal selection of network parameters. In other words, we illustrate these considerations for the real power flow, imaginary power flow, and unified power flow as the arbitrary linear combination of the real and imaginary power flows (see Chap. 6).

In this respect, a theory of power system stability [39, 40] is a potentially attractive target in the emerging deregulated power market in order to maintain efficient working of the power system while fulfilling the required demand and optimizing the economics of the power system [41]. Deregulation of the electricity market has introduced power flow uncertainty, and hence fluctuations [42, 43]. Consequently, power system planners and operators are required to run network components close to their physical capacities, under a given equilibrium operation point [39]. Uncertain power transactions cause severe voltage drops and/or disturbances between the transmission networks pertaining to the production and consumption points. In fact, the unscheduled contingencies and resulting congestions make the situation worse [40, 44], which means that a fast and effective method is required to analyze blackout conditions.

The last few decades have witnessed several major outages, and the main reasons have been causal contingencies of the lines and the associated voltage instability. In the present scenario, where there is increasing emphasis on developing an efficient power market and open access, the stability of the power system has become the main worry of both system operators and system planners [45, 46]. This means that the voltages and phases at all buses must be kept within a band of the normal operating regime, in order to run the power market effectively [47]. Due to increased demand, the continuous interconnections of nuclear, renewable, and distributed power plants to bulk power systems have led to increasingly complex operating conditions [48], with economic aspects and environmental pressures [49] creating more stringent requirements to run power systems ever closer to the stability limit while satisfying the operating constraints.

In commercial power systems, the required demand is fulfilled on the basis of the optimal power flow which minimizes the operating cost. However, if the operating point is not inside the attraction region of the equilibrium point, it is worth mentioning that this condition may be uncontrollable on the present generation scenario [41]. This can be handled only by dispatching costly generators that violate environmental constraints and/or the load curtailments, and this should be the last resort when operating a modern power system [50]. The system instability may be local or widespread and thus cause a voltage collapse, implying that post-contingency effects will be unrecoverable and highly uneconomical [51]. Due to the uncertain and varying demand, the magnitude of the voltage and the relevant phase angles bring about the voltage instability problem. With respect to the power transfer capability of the power network at physical capacity, the reactive power flow is required to maintain the voltage near an equilibrium point, especially under conditions of high load and/or certain likely fault conditions [52]. But higher reactive support reduces the capability of the network to transfer the power at its higher rating [39, 40]. In this research, we show that intrinsic geometric analysis can encode the (in)stability of power network systems.

Before going into the analysis of voltage instability, we consider that a discussion of the factors affecting these networks will be quite useful. The existing literature in the field of power system stability [39, 40, 53–57] reveals various mathematical models describing the instability of power systems. These contain both algebraic methods and differential equations. In fact, the solutions provided by these models are generally approximate for given perturbations and/or disturbances created under operation of the power system. In order to obtain a viable operating regime, the states of the system should be kept within the prescribed band with a set of precise values, whence the need for further research in this field.

We show that the differential geometric modeling approach is well suited to carry power systems instability research towards improved precision measurements against disturbances. Moreover, the approach can be extended for any network with a finite number of components, by the procedure of mathematical induction. In this case, the phenomenon is described by considering a two-bus system, in which bus-1 is treated as a generation bus feeding the required demand at bus-2 into a transmission line. The whole analysis is shown on a simplified system in order to clarify the mathematical properties and obtain a finer understanding of the proposed method. This approach may be useful in wide-area monitoring, where system state variables like real power, voltage at load buses, and phase angles are continuously observed and are prone to system disturbances.

On the basis of intrinsic geometric modeling, the operator can make operating decisions following the non-linear analysis of the considered power system. One of the main objectives of the proposed method is to provide a fast and efficient procedure which could work under the real operating conditions of power systems, allowing the operator/controller to take decisions for the viable operation of the given system. The method described in the present research encompasses the following features:

- Continuous assessment of the system state space.
- Recognition of unrecoverable system states.
- Detection of the required reactive components.

- Identification of the required reactive power.
- Implementation of the required phase correction by tap changer.
- Appropriate load shedding.
- Prediction of voltage collapse under various operating conditions.

These considerations suggest an analytical tool for system operators to access the voltage stability of a single line connected with two buses. In addition, this method can be very efficiently generalized to n-bus power systems. This problem is left open for future analysis.

Statistically, the stability of an ensemble of power systems depends not only upon the system parameters, but also on operating conditions. This introduces a high level of non-linearity into phase-space calculations. In general, the conversion of a non-linear model into the corresponding linear one can lead to significant errors in operating decisions, even for a small perturbation and/or disturbance. The instability of the power system is linked to various factors, viz., behavior of the load, reactance of transmission lines, contingencies, grid capacity, and environmental hazards, such as tsunamis, thunder storms, floods, etc. [39, 40, 49]. Thus, experience shows that power system instability is a major source of power failures around the globe, leading to uneconomical operation of such systems. It degrades the quality of the power supply and may interrupt the development of the power market as well. Continuous monitoring of power systems is essential in order to remove post-contingency effects which may be the cause of system collapse. All this shows the growing importance of the associated static and dynamic stability assessment of power systems [56, 58].

Most recent work has focused on problems such as load flow and optimal power flow, feasibility studies, and steady-state stability pertaining to the thermal limits of the apparatus. Reconfiguration of the load and generations are carried out on the lead power system operation in the limit of the near equilibrium stable point [41, 59]. There exists a vast range of publications in this direction, as attested by [58–64] and references therein. From these considerations, we anticipate that almost all instability tools are based on some approximate analysis, where factors affecting voltage instability, the effects of voltage instability, and related issues are involved. For example, Taylor series are converted into a system of linear equations by neglecting higher order terms in the relevant expansions. Eigenvalue analysis, Lyapunov's direct method, analysis of the steady-state power flow, and the Jacobian matrix approach predict the power flow limits of the network, which are indicators of a stability regime and limiting values of the parameters [60–64]. Such analyses remain a long way from the application of Hessian matrix analysis, which can incorporate higher order terms and give a more pertinent and precise solution with regard to power flow optimization, without increasing the computational burden.

Note further that regulation of the voltage can increase the power-carrying capability of the network. Analysis of voltage regulation and system instability is shown to yield an effective and useful tool for analyzing power system operation and increasing efficiency through fine-tuning of the network parameter values. With the help of the proposed intrinsic geometric procedure, impending changes in power system operation can be recognized and identified by the associated controlling parameters. It can thus help to mitigate and reduce the risk of voltage instability by proposing the underlying optimal values of the control variables. The sensitivity analysis is introduced here to enhance the effectiveness of the control variables in a given operating environment. The proposed state-space approach can further determine the natural operating point as well as the limits on the network parameters.

From the standpoint of network theory, we illustrate the parametric intrinsic geometric description by examining an exact set of pair correlation functions and the global correlation volume. Here we can take into account the inclusion of the imaginary power flow. Specifically, these considerations show that the Gaussian fluctuations about the equilibrium network components generate well-defined, nondegenerate, curved regular intrinsic Riemannian surfaces for both the purely real and the purely imaginary power flows and arbitrary linear combinations of them. In particular, consideration of [65] demonstrates that the real and imaginary power correlations involve ordinary sums of the power factors, with and without their joint effects. Analogous aspects of the intrinsic geometry analyzed in [66] yield a stable design for a class of power networks. In this context, we have shown in [66] that the intrinsic geometric model is well suited to power system planning and its operation in general. In the present model, we follow the above intrinsic geometric model, exploiting the issue of network reliability and voltage stability, and thereby examine the problem of voltage (in)stability. The robustness of the proposed model is illustrated for three important types of networks, whose exact analytical details and qualitative description are the main consideration of the present research.

In Chap. 9, regarding the role of network planning, the proposed equipment and load dynamics are the driving force behind voltage instability [67]. Thus, in a connected power system, we have illustrated the intrinsic geometry hypothesis for 2-bus systems, where the voltage of all relevant buses, i.e., both bus-1 and bus-2, varies in the same way as the transmitted power. In the sequel, we examine the state-space formulation pertaining to voltage regulation and phase-shift correction. However, our investigation does not stop there, since we analyze the voltage instability pertaining to the maximum deliverable power, i.e., for a given load. This method therefore proposes complex power optimization in a non-steady state regime by incorporating the intrinsic state-space geometry and considering the power flow between two arbitrary points along the chosen transmission line and a large voltage instability problem. In the proposed formulation, some assumptions have been made, i.e., neglect of shunt admittance, no reactive support on the load bus, the generator terminal voltage phasor is assumed to coincide with the rotor position, and the load is defined by the real and reactive power demand. In this way, we expect such an incorporation to correspond to a sub-dominant correction, and thus we leave the implementation of such issues regarding the statistical treatment for subsequent exploration.

The rest of the book has been organized as follows. In Chap. 2, we provide a brief review of some of the prerequisite concepts pertaining to network flow, the relevance of power models, and local stability, together with certain global implications arising from the basics of the real intrinsic Riemannian geometry in Chap. 3. In Chap. 4, we analyze network reliability and voltage stability problems to prepare the ground for testing these features in Chap. 5. The result proves the accuracy of this work and gives insight into the nature of stability under the transformation of the parameters of a given network configuration. In Chap. 6, we extend these considerations of geometric design and stability of power networks and examine the underlying real and imaginary phases of a network configuration. In Chaps. 7, 8, and 9, we describe a much more extensive class of exact, analytic stabilization properties from the perspective of the voltage instability problem. These are:

- voltage regulation phase-shift corrections in Chap. 7,
- voltage instability of the maximum deliverable power for a given load in the steady-state sinusoidal regime in Chap. 8, and
- stabilization of a large-scale voltage instability under the power law characterization in Chap. 9.

Finally, in Chap. 10, we present a set of conclusions and comment on the prospects for future investigations regarding the statistical aspects of network theory. Regarding the voltage (in)stability problem, we provide explicit mathematical formulae in the Appendices.

References

1. R. Billington, R.J. Ringle, A.J. Wood, *Power-System Reliability Calculations* (MIT Press Classics Series, 1973) 173p, ISBN-10: 026202098X
2. A. Chakraborty, P. Sen, An analytical investigation of voltage stability of an EHV transmission network based on load flow analysis. J. Inst. Eng. (India) Electr. Eng. Div. **76** (1995)
3. P. Kundur, *Power System Stability and Control*, EPRI Power System Engineering Series (McGraw-Hill, New York, 1994), p. 328
4. H. Frank, B. Landstorm, Power factor correction with thyristor-controlled capacitors. ASEA J. **45**, 180–184 (1971)
5. R. Rajarman, F. Alvarado, A. Maniaci, R. Camfield, S. Jalali, Determination of location and amount of series compensation to increase power transfer capability. IEEE Trans. Power Syst. **13**(2), 294–300 (1998)
6. H. Almasoud, Shunt capacitance for a practical 380 kV system. Int. J. Electr. Comput. Sci. **9**(10), 23–27 (2009)
7. G. Bonnard, The problem posed by electrical power supply to industrial installations. Proc. IEEE Part B **132**, 335–340 (1985)
8. K. Ramalingam, C.S. Indulkar, Transmission line performance with voltage sensitive loads. Int. J. Electr. Eng. Educ. **41**(1), 64–70 (2004)
9. H. Frank, S. Ivner, Thyristor controlled shunt compensation in power networks. ASEA J. **54**, 121–127 (1981)
10. H.-T. Nguyen Vu, Voltage control with shunt capacitance on radial distribution line with high R/X factor Electrical Engineering, Polytechnic University of HCMC, 2005
11. B. Milosevic, M. Begovic, Capacitor placement for conservative voltage reduction on distribution feeders. IEEE Trans. Power Deliv. **193**, 1360–1367 (2004)
12. E.H. Camm, Shunt capacitor over voltages and a reduction technique, in *IEEE/PES, Transmission and Distribution Conference and Exposition*, New Orland, LA, (1999)
13. G. Fusco, *Adaptive Voltage Control in Power Systems: Modeling, Design, and Applications*, Advances in Industrial Control (Springer, London, 2006). ISBN: 84628564X
14. Canadian Electrical Association, *Static Compensation for Reactive Power Control* (Context Publications, Winnipeg, 1984)
15. P.M. Anderson, R.G. Farmer, *Series Compensation of Power Systems* (Fred Laughter and PBLSH Inc., Encinitas, 1996) ISBN-10: 1888747013

16. J. Stones, A. Collinson, Introduction to power quality. Power Eng. J. **15**(2), 58–64 (2001)
17. T.J. Miller, *Reactive Power Control in Electrical Systems* (Wiley, New York, 1982)
18. G. Radman, R.S. Raje, Power flow model/calculation for power system with multiple FACTS controllers. Electr. Power Syst. Res. (Elsevier, ScienceDirect) **77**, 1521–1531 (2007)
19. T.J.E. Miller (ed.), *Reactance Power Control in Electric Systems* (Wiley, New York, 1982)
20. M.H. Shwedhi, M.R. Sultan, Power factor correction capacitors: essentials and cautions. IEEE Power Eng. Soc. Summer Meet. **3**, 1317–1322 (2000)
21. R.T. Saleh, A.E. Emanuel, Optimum shunt capacitor for power factor correction at busses with high distorted voltage. IEEE Trans. Power Deliv. PWRD **2**(1), 165–173 (1987)
22. C.-T. Chi, Evaluation of performance of a novel voltage compensation strategy for an ac contactor during voltage sags. Int. J. Innov. Comput. Inf. Control **4**(11), 2809–2822 (2008)
23. A.R. Sefie, A new hybrid optimization method for optimum distribution capacitor planning. Mod. Appl. Sci. **3**(4), 196–202 (2009)
24. N. Ng, M.A. Salama, A.Y. Chikhani, Classification of capacitor allocation techniques. IEEE Trans. Power Deliv. **15**(1), 387–392 (2000)
25. T.V. Cutsem, V.H. Quintana, Network parameter estimation using online data with application to transformer tap position estimation. Gener. Transm. Distrib. IEEE Proc. C **135**(1), 31–40 (2006)
26. Q.-P. Zhang, C.-M. Wang, Z.-J. Hou, Power network parameter estimation method based on data mining technology. J. Shanghai Jiaotong Univ. (Sci.) **13**(4), 468–472 (2008)
27. Powerworld Simulator, Available from the homepage of Power World Corporation at website, www.powerworld.com (2012)
28. O. Anaya-Lara, E. Acha, Modeling and analysis of custom power systems by PSCAD/EMTDC. IEEE Trans. Power Deliv. PWDR **17**(1), 266–272 (2002)
29. G. Ruppeiner, Riemannian geometry in thermodynamic fluctuation theory. Rev. Mod. Phys. **67**(3), 605–659 (1995)
30. B.N. Tiwari, *Sur les corrections de la géométrie thermodynamique destrous noirs*, Éditions Universitaires Européennes, Germany (2011). ISBN: 978-613-1-53539-0; arXiv:0801.4087v2 [hep-th]
31. S. Bellucci, B.N. Tiwari, On the microscopic perspective of black branes thermodynamic geometry. Entropy **12**, 2096 (2010); arXiv:0808.3921v1
32. S. Bellucci, B.N. Tiwari, An exact fluctuating 1/2 BPS configuration. Springer, J. High Energy Phys. **05**, 23 (2010), arXiv:0910.5314v1
33. J. Grainger Jr., W. Stevenson, *Power System Analysis*, 1st edn. (McGraw-Hill Science, Engineering, Math, New York, 1994)
34. G. Ruppeiner, Riemannian geometry in thermodynamic fluctuation theory. Rev. Mod. Phys. **67**, 605 (1995) [Erratum **68**, 313 (1996)]
35. G. Ruppeiner, Thermodynamics: a riemannian geometric model. Phys. Rev. A **20**, 1608 (1979)
36. G. Ruppeiner, Thermodynamic critical fluctuation theory? Phys. Rev. Lett. **50**, 287 (1983)
37. G. Ruppeiner, New thermodynamic fluctuation theory using path integrals. Phys. Rev. A **27**, 1116 (1983)
38. G. Ruppeiner, C. Davis, Thermodynamic curvature of the multicomponent ideal gas. Phys. Rev. A **41**, 2200 (1990)
39. M. Crappe, *Electric Power Systems* (ISTE Ltd, New York, 2008)
40. A.R. Bergen, V. Vittal, *Power System Analysis*, 2nd edn. (Prentice Hall, Upper Saddle River, 2000)
41. D.P. Brook, R.W. Dunn, Improving power system stability and economy by coordination of controller settings and power constraints. *Power Engineering Society Winter Meeting* (2001). IEEE vol. 2, pp. 499–503 (2001)
42. A.K. Al-Othman, M.R. Irving, Uncertainty modeling in power system state estimation. IEEE Proc. Gener. Transm. Distrib. **152**(2), 233–239 (2005)
43. D. Wu, H. Xin, D. Gan, Evaluating the impact of uncertain parameters in power system dynamic simulations. Electr. Power Syst. Res. **78**(11), 1965–1971 (2008)

44. M. Esmaili, H.A. Shayanfar, N. Amjady, Congestion management enhancing transient stability of power systems. Appl. Energy **87**(3), 971–981 (2010)
45. V. Ajjarapu, C. Christy, The continuation power flow: a tool for steady state voltage stability analysis. IEEE Trans. Power Syst. **7**, 416–423 (1992)
46. A.J. Calvaer, Voltage stability and collapses: a simple theory based on real and reactive currents. Revue Générale de l'Électricité **8**, 1–17 (1986)
47. T.V. Cutsem et al., Determination of secure operating limits with respect to voltage collapse. IEEE Trans. Power Syst. **14**(1), 327–333 (1999)
48. F. Murphy, Y. Smeers, Capacity expansion in imperfectly competitive restructured electricity market. Oper. Res. **53**, 4 (2002)
49. Technical Report, Environmental and health impacts of electricity generation. International Energy Agency, *Implementing Agreement for Hydropower Technologies and Programmes* (2002)
50. C.W. Taylor, Concept of undervoltage load shedding for voltage stability. IEEE Trans. Power Deliv. **7**, 480–488 (1992)
51. J. Ma, Advanced techniques for power system stability analysis, Ph.D. Thesis, School of Information Technology and Electrical Engineering, University of Queensland, 2008
52. H.K. Clark, Voltage control and reactive supply problems, in *IEEE Tutorial Course on Reactive Power: Basics, Problems and Solutions*, Publication 87 EH0262-6-PWR, presented at the IEEE-PES Summer Meeting, San Francisco, CA, July 12–17, 1987, and the Winter Meeting, New York, NY, 1988
53. IEEE Working Group on Voltage Stability, Suggested techniques for voltage stability analysis, IEEE Publication PWR **5**, 93TH0620 (1993)
54. J. Bian, P. Rastgoufard, Power system voltage stability and security assessment. Electr. Power Syst. Res. **30**(3), 197–200 (1994)
55. M. Ribbens-Pavella, F.J. Evans, Direct methods for studying dynamics of large-scale electric power systems: a survey. Automatica **21**(1), 1–21 (1985)
56. L.D. Colvara, Stability analysis of power systems described with detailed models by automatic method. Int. J. Electr. Power Energy Syst. **31**(4), 139–145 (2009)
57. B.I. Lima Lopes, A.C. Zambroni de Souza, A Newton approach for long term stability studies in power systems. Appl. Math. Comput. **215**(9), 3327–3334 (2010)
58. P.-A. Löf, D.J. Hill, S. Arnborg, G. Andersson, On the analysis of long-term voltage stability. Inter. J. Electr. Power Energy Syst. **15**(4), 229–237 (1993)
59. M.A. Kashem, V. Ganapathy, G.B. Selengor, Network reconfiguration for enhancement of voltage stability in distribution networks. IEEE Proc. Gener. Transm. Distrib. **147**(3), 171–175 (2000)
60. J.D. Glover, M.S. Sarma, *Power Systems Analysis and Design*, 3rd edn. (Brooks/Cole, Pacific Grove, 2002)
61. H. Kwatny, L. Bahar, A. Pasrija, Energy-like Lyapunov functions for power system stability analysis. IEEE Trans. Circuits Syst. **32**(1), 1140–1149 (1985)
62. Y.Z. Sun, X. Li, M. Zhao, Y.H. Song, New Lyapunov function for transient stability analysis and control of power systems with excitation control. Electr. Power Syst. Res. **57**(2), 123–131 (2001)
63. K.S. Chandrashekhar, D.J. Hill, Cutset stability criterion for power systems using a structure-preserving model. Int. J. Electr. Power Energy Syst. **8**(3), 146–157 (1986)
64. J.E.O. Pessanha, O.R. Saavedra, J.C.R. Buzar, A.A. Paz, C.P. Poma, Power system stability reinforcement based on network expansion: a practical case. Int. J. Electr. Power Energy Syst. **29**(3), 208–216 (2007)
65. N. Gupta, B.N. Tiwari, S. Bellucci, Geometric design and stability of power networks, arXiv:1011.2924 [stat.AP]
66. N. Gupta, B.N. Tiwari, S. Bellucci, Intrinsic geometric analysis of network reliability and voltage stability, arXiv:1011.2929 [stat.AP]
67. B.N. Tiwari, N. Gupta, S. Bellucci, State-space perspective to voltage instability. Inter. J. Control Autom., (submitted)

Chapter 2
Proposed Methodology

In this chapter, we recall the mathematical preliminaries and the flow properties of power networks. The method used here is to eliminate the effect of voltage fluctuation about an equilibrium. In this process, we determine the required optimum values of the underlying parameters, such as inductance L and capacitance C, along with any others. We shall also set up a method to predict the exact values of R and L in order to increase the reliability and efficiency of the network. From the described criteria, one can opt for single circuit or double circuit lines between two buses to ensure the reliability of the network. All prior conventional solutions use load flow equations and characteristics and performance equations of transmission lines (see, for instance, [1, 2]). In general, the power conservation equations associated with the real (resistive) and imaginary (reactive) branch parameters are given respectively by

$$P_i = \sum_j |V_i||V_j||Y_{ij}|(G_{ij} \cos \theta_{ij} + B_{ij} \sin \theta_{ij}) \tag{2.1}$$

and

$$Q_i = \sum_j |V_i||V_j||Y_{ij}|(G_{ij} \sin \theta_{ij} - B_{ij} \cos \theta_{ij}), \tag{2.2}$$

where V_i and V_j are the voltages corresponding to the ith and jth buses, respectively (see Refs. [1, 2] for an introduction to power flow models/calculations and power systems with multiple FACTS controllers and associated analysis of the power system networks). In the above equations, the phases θ_{ij} are defined by

$$\tan \theta_{ij} = X_{L_{ij}} - X_{C_{ij}} r_{ij}. \tag{2.3}$$

Using J as the imaginary unit, we define the impedance by

$$Y_{ij} = \frac{1}{Z_{ij}} = \frac{1}{r_{ij} + J(X_{L_{ij}} - X_{C_{ij}})}, \tag{2.4}$$

S. Bellucci et al., *Geometrical Methods for Power Network Analysis*,
SpringerBriefs in Electrical and Computer Engineering,
DOI: 10.1007/978-3-642-33344-6_2, © The Author(s) 2013

and the phase for the jth bus by

$$\delta_j = \frac{V_j}{|V_j|}. \tag{2.5}$$

Thus, the steady-state condition $|V_i| = 1$ for the ith bus describes the equilibrium configuration under voltage fluctuations. In the above cases, we are interested in the following standard network considerations:

- A lossless line is defined by $\theta_{ij} = \pm 90$ with relative phases δ_i, where

$$\delta_i - \delta_j \leq \frac{\pi}{6}. \tag{2.6}$$

- For the real power flow with the resistances $\{r_i\}$, the θ_{ij} are defined by the expression

$$\tan \theta_{ij} := \frac{X_{L_{ij}}}{r_{ij}}. \tag{2.7}$$

- In general, when the reactive power flow is allowed to possess a nonzero voltage fluctuation, we define θ_{ij} by

$$\tan \theta_{ij} = \frac{X_{L_{ij}} - X_{C_{ij}}}{r_{ij}}. \tag{2.8}$$

At this juncture, we propose to define the local variations of the loads by the Hessian matrix of the total effective power in the network or at the chosen component. The robustness of the proposed model is illustrated by introducing variations in the chosen circuit parameters and load. We exploit the fact that the power fluctuations can be described by the critical exponent of the correlation equations. Note that this model is able to predict the condition and state of every branch of the network, and thus whether it is robust from the perspective of power system planning.

Regarding the network model specification, the notion of power flow in the Riemannian geometry framework supports the proposal of an intrinsic network reliability [3, 4] and global voltage stability [5] in the case of an abrupt load change at any disturbed bus of the power system. To begin with the novel feature of the present research, we propose an admissible characterization of the network variables.

2.1 Proposition

The impedance variables of the power network form an admissible basis.

Proof We demonstrate the above proposition by performing a coordinate transformation. To do so, let $\{L, R, C\}$ be a set of mutually independent effective parameters of the network. Let us consider the following coordinate transformations on the LCR configuration

$$(r, X_L, X_C) = (R, \omega L, 1/\omega C). \tag{2.9}$$

The Jacobian of the transformation is thus given by

$$J(\omega, C) = \begin{pmatrix} 1 & 0 & 0 \\ 0 & \omega & 0 \\ 0 & 0 & -1/\omega C^2 \end{pmatrix}. \tag{2.10}$$

As a function of $\{\omega, C\}$, the determinant $|J|$ of the Jacobian matrix is

$$|J| = -\frac{1}{C^2}. \tag{2.11}$$

We thus observe that this provides a sort of fine-tuning for the voltage fluctuations. It shows that the voltage fluctuation problem can be solved by a variable capacitor along with variations of the other parameters. Consequently, we can analyze LCR fluctuations either in the impedance basis or in the basic component parameter basis, as long as the fluctuating component has a nonzero capacitance, i.e., $C \neq 0$. Equivalently, both of the above characterizations describe the same configuration of chosen component(s) and thus the whole network. For a nonzero base frequency, it is worth mentioning that the proposition holds for the LR component. This follows directly from the fact that the determinant of the Jacobian takes the above value.

Let us consider some salient features of our network characterizations. For the power network system, our model specification shows that there exists a set of constant voltages in equilibrium. Therefore, the corresponding phases between the reactive components are given by the extrema of the total power. The basis vectors are represented by the points on the intrinsic state-space manifold [6–9].

Considering the theory of random variations along with the laws of equilibrium circuit configurations, $\{x_i\}$ leads to Riemannian geometric structures [9]. Over an equilibrium basis, defined as a finite set of the parameters

$$x_i = \{X_{L_i}, X_{C_i}, r_i\}, \tag{2.12}$$

they form the coordinate charts on the reactive configuration. Henceforth, the states of interest will be understood as the finite collection of points $\{x_i\}$.

Following [9], the present analysis establishes that the invariant distance between two arbitrary equilibrium states is inversely proportional to the random variation connecting the two states. In particular, a 'less probable variation' signifies that the states are 'far apart'.

2.2 Admissible Choice of Component(s)

In this model, our analysis utilizes knowledge of the above real and complex power flow equations. We thereby determine the nature of local and global reliability under the thermal limit of the network, and voltage stability for uncertain voltage and hence fluctuations at the disturbed buses. Using parallel lines, we predict the required augmentation for the network reliability and required value of C to keep the voltage level under the stability limit in current scenarios of power flow considerations. As a consequence of our model, one can determine the intrinsic geometric nature of the power system for a given set of network parameters.

2.3 Theoretical Motivation

The model under consideration is built from the viewpoint of intrinsic power system planning and operation in modern applications. According to existing electrical network designs, we identify the following cases as interesting from the planning viewpoint. Correspondingly, we illustrate the hypothesis of the intrinsic geometric surface type of power flow model shown in the flow chart of Fig. 2.1, for the parameters in the components, disturbed buses, and generators of the network configuration. To determine the reliability of the network, we consider the loss in power flow, which modifies the set of inductive variables to the combined set of resistances and inductances $\{X_{L_i}\}$, while the voltage fluctuation problem is investigated by including the capacitances in the above set of variables $\{r_i + X_{L_i}\}$. Thus, the union of the sets $\{r_i + X_{L_i}\}$ for the RL network system and $\{r_i + X_{L_i} + X_{C_i}\}$ for the LCR network system forms the basis vectors on the manifold M_n. In particular, the set of parameters $\{x_1, \ldots, x_n\}$ offers a combined intrinsic geometric analysis for network reliability and voltage stability.

According to the above notion of fluctuation theory (see, for example, [10]), a fluctuating power network reaches a local equilibrium when the network flow parameters $\{x_1, \ldots, x_n\}$ are held fixed. In order to implement this idea, let us consider a two-parameter power system stabilization whose fluctuation properties we wish to compute. As a function of the given parameters $\{x_1, x_2\}$, the power $P(x_1, x_2)$ may be represented by the map

$$P : \mathcal{M}_2 \to R, \tag{2.13}$$

by assigning the set theoretic rule $(x_1, x_2) \mapsto P(x_1, x_2)$ from the local power surface $\mathcal{M}_2$ to the set of real numbers $\mathbb{R}$. In order to obtain such a fixed point analysis of the power $P(x_1, x_2) \in \mathbb{R}$, we may choose a fixed point $(x_1^0, x_2^0) \in \mathcal{M}_2$ and expand the power $P(x_1, x_2)$ in a Taylor series about the chosen fixed point (x_1^0, x_2^0):

Fig. 2.1 The intrinsic geometric hypothesis represented by the flow chart for the power flow model, describing fluctuations in the parameters of the components, disturbed buses, and generator of the network configuration

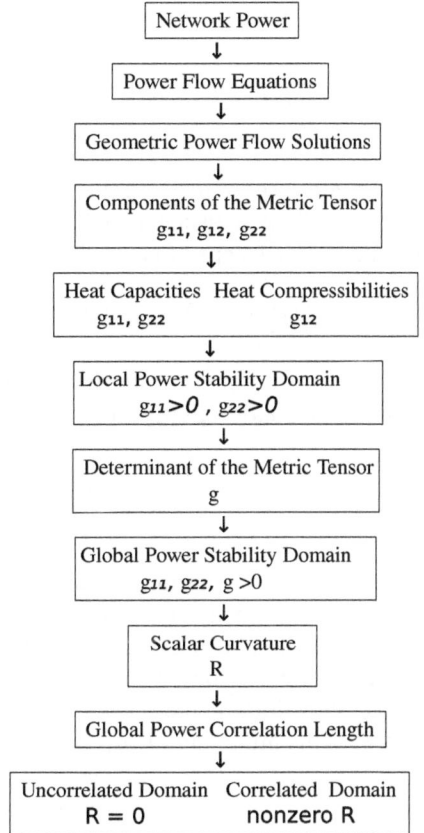

$$P(x_1, x_2) = P(x_1^0, x_2^0) + \frac{\partial P}{\partial x_1}\bigg|_{(x_1^0, x_2^0)} (x_1 - x_1^0) + \frac{\partial P}{\partial x_2}\bigg|_{(x_1^0, x_2^0)} (x_2 - x_2^0)$$

$$+ \frac{\partial^2 P}{\partial x_1^2}\bigg|_{(x_1^0, x_2^0)} (x_1 - x_1^0)^2 + 2 \frac{\partial^2 P}{\partial x_1 \partial x_2}\bigg|_{(x_1^0, x_2^0)} (x_1 - x_1^0)(x_2 - x_2^0)$$

$$+ \frac{\partial^2 P}{\partial x_2^2}\bigg|_{(x_1^0, x_2^0)} (x_2 - x_2^0)^2 + \cdots . \tag{2.14}$$

The fixed points (x_{1i}^0, x_{2i}^0) of the power $P(x_1, x_2)$ can be determined by solving the flow equations

$$\frac{\partial P}{\partial x_1} = 0, \qquad \frac{\partial P}{\partial x_2} = 0. \tag{2.15}$$

For a given power $P(x_1, x_2)$, this defines the fixed point set

$$S_P := \left\{ (x_1^i, x_2^i) \in \mathcal{M}_2 \; \middle| \; \frac{\partial P}{\partial x_1} \bigg|_{(x_1^i, x_2^i)} = 0, \quad \frac{\partial P}{\partial x_2} \bigg|_{(x_1^i, x_2^i)} = 0, \quad i \in \Lambda \right\}, \quad (2.16)$$

where Λ is a set of finite cardinality. In the thermodynamic limit, we shall consider that $\Lambda \to \mathbb{R}$. We consider that the chosen power system evolves through an ensemble of infinitesimal steps of network parameter fluctuations on the lattice:

$$(x_1^0, x_2^0), \; (x_1^0 + h, x_2^0 + h), \; (x_1^0 + 2h, x_2^0 + 2h), \ldots, \; (x_1^0 + nh, x_2^0 + nh). \quad (2.17)$$

Then we may represent the local fixed points of the network system by a two-dimensional lattice giving a two-dimensional data system

$$
\begin{aligned}
x_1^a &= x_1^0 + ah, \quad a = 0, 1, 2, \ldots, n, \\
x_2^b &= x_2^0 + bh, \quad b = 0, 1, 2, \ldots, n.
\end{aligned}
\quad (2.18)
$$

Through the above formulation, we may obtain a discretization of the power surface $\mathcal{M}_2$ as a two-dimensional random data system. It is worth mentioning that we may generalize the above evolution of the data theoretic setup for the chosen power system using a distinct pair (h_1, h_2) as the step sizes of the fluctuations in the x_1 and x_2 directions on the surface $\mathcal{M}_2$. In this case, we may introduce a new step size $h \in \mathbb{R}$ as the minimum of $(h_1, h_2) \in \mathbb{R}^2$, i.e.,

$$h := \min(h_1, h_2). \quad (2.19)$$

Statistically, with the above convention, we can add a finite number of extra points into the data sample. In particular, by considering the central limit theorem, we can take into account the Gaussian distributions of the system parameters. On the other hand, let $h = \max(h_1, h_2)$ be the maximum of the sizes $(h_1, h_2) \in R^2$. Then, by excluding a finite number of data points from the sample, the application of the central limit theorem yields the same Gaussian distribution of the network parameters of the fluctuating power system in the continuous limit. In such a process of network equilibrium formation, the distribution theory of the power network, the error tends to a similar limit in the discrete and continuous limits of the chosen power system. In the sequel, and in particular in Sect. 3.2, we shall present the above-mentioned geometric framework from the fluctuation theory setup, giving an intrinsic Riemannian manifold M_n of parameters for a given power network configuration.

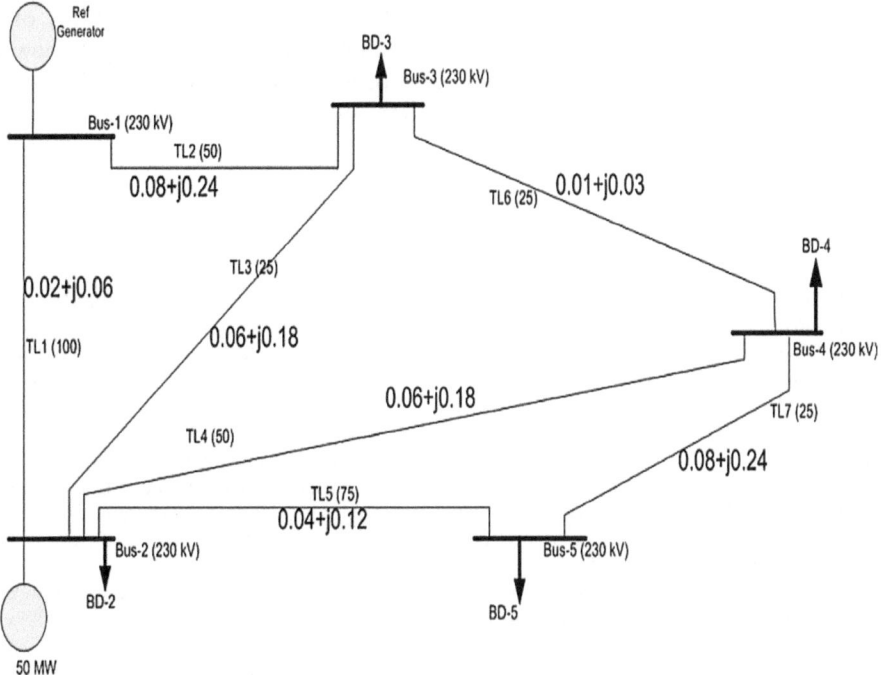

Fig. 2.2 An IEEE 5-bus circuit, showing the fluctuations of the parameters in the components, disturbed buses, and generator of a network configuration

2.4 Circuit Planning

In this innovation, the useful features of the proposed model, network reliability, and voltage stability of the power networks are described for any circuit. This is exemplified by the IEEE 5-bus network shown in Fig. 2.2. This model can be applied to a broad class of high voltage networks as well. The line parameters $\{r_i + X_{L_i}\}$ are given on the respective transmission lines.

2.5 Validity of the Proposed Model

Since the proposed model is a non-linear improvement on existing linear models, this ensures that the intrinsic geometric analysis will offer the best network reliability and voltage stability for non-linear fluctuations of the real and complex power flow across network component(s). Consequently, we examine the test of transmission line reliability by considering an LR component of the power flow. This follows from consideration of the real power flow equation (2.1), without considering

C compensation. Similarly, we demonstrate that the test for voltage stability of a transmission line follows from joint consideration of the real and complex power flow equations (2.1) and (2.2). In this context, specific considerations concerning LR network and LCR network systems are given in subsequent chapters.

References

1. G. Radman, R.S. Raje, Power flow model/calculation for power system with multiple FACTS controllers. Electr. Power Syst. Res. (Elsevier, ScienceDirect) **77**, 1521–1531 (2007)
2. J. Grainger Jr., W. Stevenson, *Power System Analysis*, 1st edn. (McGraw-Hill Science, Engineering, Math, New York, 1994)
3. G.A. Mass, Reliability issues of the UCTE systems. *IEA Workshop on Transmission Reliability in Competitive Electricity Markets*, Paris, March, 2004, pp. 29–30
4. U.S., Canada Power System Outage Taskforce: Interim report: Cause of the August 14th blackout in the United States and Canada, North America Electricity Reliability Council, Nov (2003)
5. E. Irving, J.P. Barret, C. Charcossey, J.P. Monville, Improving power network stability and unit stress with adaptive generator control. Automatica **15**(1), 31–46 (1979); (Available online 30 January 2003 from ScienceDirect)
6. B.N. Tiwari, *Sur les corrections de la géométrie thermodynamique des trous noirs*, (Éditions Universitaires Européennes, Germany (2011). ISBN 978-613-1-53539-0; arXiv:0801.4087v2 [hep-th]
7. S. Bellucci, B.N. Tiwari, On the microscopic perspective of black branes thermodynamic geometry. Entropy **12**, 2096 (2010); arXiv:0808.3921v1
8. S. Bellucci, B.N. Tiwari, An exact fluctuating 1/2 BPS configuration. Springer J. High Energy Phys. **05**, 23 (2010); arXiv:0910.5314v1
9. G. Ruppeiner, Riemannian geometry in thermodynamic fluctuation theory. Rev. Mod. Phys. **67**(3), 605–659 (1995)
10. G. Ruppeiner, Riemannian geometry in thermodynamic fluctuation theory. Rev. Mod. Phys. **67**, 605 (1995); [Erratum **68**, 313 (1996)]

Chapter 3
Intrinsic Geometric Characterization

In this chapter, we offer a concise account of network power flow and stability criteria arising from real intrinsic Riemannian geometry. We begin by considering a brief review of the flow equations and related concepts, for use in the later chapters. The general details of the network theory considered in this research can be found in the Refs. [1–6].

3.1 Origin of Thermodynamic Geometry

In this section, we offer a brief outline concerning the origin and theory of fluctuations in statistical mechanics. Subsequently, we shall apply the above theory to two- and three-parameter fluctuating power network systems.

Ruppeiner used the physics of black holes to reinterpret the assumptions of power network systems introduced in the last chapter [7]. For a given black hole ensemble, the statement that all the statistical degrees of freedom of a black hole live on the black hole event horizon can be taken as an indication that the state-space scalar curvature gives the average number of correlated Planck areas on the event horizon of the black hole. In this context, there are many such considerations involving Riemannian geometric models, unifying the thermodynamic curvature and phase transitions, critical fluctuation theory, path integral properties, ideal gases, black holes, uncertainties, the physics of correlation, and information geometries, these being exemplified in [8–27]. Furthermore, it is worth mentioning the importance of Legendre associated configurations pertaining to the Legendre-transformed chemical geometry of equilibrium thermodynamics [28–32].

Here we are motivated by the ideas of intrinsic geometry introduced in Chap. 2, and in particular the Taylor series expansion of the power given in (2.14). Since it has been observed for configurations involving black holes in string theory [33–39] and M-theory [40–43] that they possess a rich set of stability structures, we shall specialize our present considerations for the fluctuating power network systems. In this respect, we have previously focused on the importance of intrinsic geometry in [13, 14, 16, 18–23] for various black hole configurations. Indeed, the intrinsic geometry approach has been implemented in the past for carrying out the stability analysis of black hole solutions in general relativity [24–27], attractor black holes [44–48], and Legendre-transformed finite-parameter chemical configurations [28, 29], quantum field theory and associated hot QCD backgrounds [30, 31], and the stability properties of quarkonium states [32]. As mentioned in Chap. 2, following the notion of Ruppeiner geometry [8], we shall discuss the equilibrium perspective of power network statistical configurations, explaining the nature of local pair correlations, global ensemble stabilities, and the associated global geometric stability.

3.2 Network Fluctuation Theory

The above-mentioned general notions have been analyzed in the context of solid state physics. See, for instance, [7, 9–12, 49, 50], where phase transitions and free energy analysis have been considered for the local and global stability of condensed matter systems, such as ideal gases, and electric and magnetic systems. In the same spirit, we now consider flow properties and stability characterization for electrical networks. We thus define the underlying fluctuations about a given equilibrium in terms of the network parameters.

As for the intrinsic geometric considerations mentioned in [7], the present analysis provides the parametric pair correlations and their correlation relation functions about a given equilibrium statistical configuration. We then consider a network with a finite set of parameters $\{x_i\}_{i=1}^n$ which have equilibrium values $\{x_{i0}\}_{i=1}^n$. Notice that the network parameters $\{x_i\}_{i=1}^n$ may in general be complex-valued. With an appropriate normalization convention, the total network power $P(x_i)$ can be considered as a real-valued function by expressing each parameter x_i as a set of real pairs:

$$x_i = (x_{i1}, x_{i2}), \quad i = 1, 2, \ldots, n . \tag{3.1}$$

This doubles the total number of parameters and introduces the real stabilization framework for characterizing the network flow problem. Thus, the flow issues under consideration may be analyzed by extremizing the total power $P(x_i)$. As mentioned in Chaps. 1 and 2, and in particular (2.16) for the case of two-parameter networks, extremization of a general power $P(x_i)$ gives the following set of critical points

$$\mathscr{C} := \{x_i^0 |\, i = 1, 2, \ldots, 2n\} . \tag{3.2}$$

This is easily determined by the criticality condition

$$\partial P(x_i) = 0 . \tag{3.3}$$

It is worth noting that the stability of the underlying network configuration is straight-forwardly achieved by demanding the positivity of the Hessian matrix of the power $P(x_i)$. This is exemplified by (2.14) for the particular case of a two-parameter power network system. We also note that the ordinary Hessian matrix $\partial_i \partial_j P(x_i)$ defines a symmetric bilinear form and thus provides a real intrinsic metric tensor g_{ij} defined by

$$g_{ij} = \partial_i \partial_j P(x_i)\big|_{x_i = x_i^0} . \tag{3.4}$$

This means that the stability analysis of the given network system may be carried out by examining the positivity of the principal minors of the covariant metric tensor g_{ij}.

The reader is referred to [7–14, 16, 18–23] for detailed applications of intrinsic geometry and associated considerations for network engineering. Note that the notion of intrinsic geometry as stated above continues to hold in general for any network with finitely many parameters. Following the above hypothesis, we provide a brief review of the intrinsic geometry for the two-parameter network configuration and illustrate the underlying parametric correlation properties.

3.3 Power Flow Fluctuations

To be specific, we illustrate the idea of state-space geometry for the case of two-parameter configurations, defined by parameters $\{x_1, x_2\}$. Let $S(x_1, x_2)$ be the complex network power. For a given $S(x_1, x_2)$, the components of the correlation functions are described by the Hessian matrix $\text{Hess}(S(x_1, x_2))$ of the generalized power function under the flow of the parameters. The components of the intrinsic metric tensor are then

$$g_{x_1 x_1} = \frac{\partial^2 S}{\partial x_1^2} , \quad g_{x_1 x_2} = \frac{\partial^2 S}{\partial x_1 \partial x_2} , \quad g_{x_2 x_2} = \frac{\partial^2 S}{\partial x_2^2} . \tag{3.5}$$

The components of the intrinsic metric tensor are associated with the pair correlation functions of the given complex power flow. It is worth mentioning that the coordinates of the underlying power lie on the surface of the parameters $\{x_1, x_2\}$, which in the statistical sense gives the origin of the fluctuations in the network. This is because the components of the metric tensor comprise the Gaussian fluctuations of the network power, which is a function of the parameters of the power configuration. For a given power network, the local stability of the underlying system requires both the principal components to be positive. The diagonal components of the metric tensor $\{g_{x_i x_i} \mid i \in 1, 2\}$ correspond to the heat capacities of the system, and are thus required to be positive-definite quantities:

$$g_{x_i x_i} > 0 , \quad i = 1, 2 . \tag{3.6}$$

From the perspective of intrinsic geometry, the stability properties of the network flows can thus be obtained from the positivity of the determinant of the metric tensor. For the Gaussian fluctuations of the two-charge equilibrium power configurations, the existence of a positive-definite volume form on the power surface imposes such a stability condition. Thus, a power supplying configuration is said to be stable if the determinant

$$\|g\| = S_{x_1 x_1} S_{x_2 x_2} - S_{x_1 x_2}^2 \tag{3.7}$$

of the metric tensor remains positive. For two-parameter networks, the geometric quantities corresponding to the chosen power elucidate the typical features of the Gaussian fluctuations about an ensemble of equilibrium configurations. As a global invariant, the intrinsic scalar curvature provides information about the correlation volume of the underlying power fluctuations. Explicitly, the scalar curvature R is given by

$$
\begin{aligned}
R = -\frac{1}{2(S_{x_1 x_1} S_{x_2 x_2} - S_{x_1 x_2}^2)^2} \Big(&S_{x_2 x_2} S_{x_1 x_1 x_1} S_{x_1 x_2 x_2} + S_{x_1 x_2} S_{x_1 x_1 x_2} S_{x_1 x_2 x_2} \\
&+ S_{x_1 x_1} S_{x_1 x_1 x_2} S_{x_2 x_2 x_2} + S_{x_1 x_2} S_{x_1 x_1 x_1} S_{x_2 x_2 x_2} \\
&- S_{x_1 x_1} S_{x_1 x_2 x_2}^2 - S_{x_2 x_2} S_{x_1 x_1 x_2}^2 \Big) .
\end{aligned}
\tag{3.8}
$$

Note that zero scalar curvature indicates that the power of the network fluctuates independently of the flow parameters, while a divergent scalar curvature signifies a sort of phase transition, indicating an ensemble of highly correlated pixels of information on the power surface. In the following, e.g., in (4.14) and (7.7), we exemplify the above remarks in the case of vanishing Ricci scalar curvature. For examples of an intrinsic scalar curvature which can exhibit divergences, see for instance (5.17) for the LCR network and (6.22) for real power flow optimization.

For the case of a two-parameter network, the above analysis of the surface shows that the scalar curvature and Riemann curvature tensor satisfy

$$R(x_1, x_2) = \frac{2}{\|g\|} R_{x_1 x_2 x_1 x_2} . \tag{3.9}$$

The scalar curvature defined thus informs as to the nature of the long-range global correlations and underlying phase transitions originating from the power flow. In this sense, we suggest that an ensemble of signals corresponding to the network are statistically interacting if the underlying power configuration has a nonzero scalar curvature. Indeed, we may also note that the configurations considered here are allowed to be effectively attractive or repulsive, and weakly interacting. In general, the intrinsic geometric analysis also provides a set of physical indications encoded in the geometrically invariant scalar curvature. For the electrical network, the underlying analysis would involve an ensemble or subensemble of the equilibrium configuration forming a statistical basis about the Gaussian distribution.

As far as the physical aspects of power networks are concerned, it turns out that the network stabilization problem can be extended to a number of electrical configurations, including non-steady regimes and large vacuum moduli disturbances. In fact, the detailed computations and qualitative depictions of the underlying intrinsic geometric quantities take us to the forefront of modern network theory. In order to realise the above scenarios, the present approach depends on the number of fluctuating network parameters. By considering a given set of such parameters, we can provide a detailed characterization of the network stabilization problem through the notion of real intrinsic Riemannian geometry. In fact, we have extended the state-space analysis for the non-steady state regimes and anticipated the characterization of arbitrary network (in)stabilities.

3.4 Stability of Minimally Coupled Buses

From the above considerations, if there are n minimally coupled transmission lines connected to a given bus, then the stability of that bus is defined by the metric tensor

$$A_{yz} = \text{diag}(g_{ij(n)}) \,, \tag{3.10}$$

where $A_{yz} = \partial_y \partial_z S(x^1, x^2, \ldots, x^n)$ and $y, z = 1, 2, \ldots$. In the above setup, the elements appear as the diagonal elements of the total system. Thus, all the principal minors remain equal to or greater than zero for a stable voltage profile. The determinant $\det g_{ij(n)}$ depends on the foregoing determinants $\det g_{ij(n-1)}$ for all n. This characterization improves the performance of the power system in situations of sudden load change or transmission line outage redundancy.

3.5 Intrinsic Network Stabilization

As mentioned in the last section, it follows that linear stability may simply be obtained by demanding the positivity of the principal components of the real metric tensor. Thus, the system is stable along an intended dimension n if the corresponding component satisfies $g_{ij} > 0$. Furthermore, it follows that the configuration is stable on the two-dimensional surface defined by the coordinate chart $\{x_1, x_2\}$ of the full network if the relevant determinant of the metric tensor satisfies

$$g_2 = g_{11}g_{22} - g_{12}^2 > 0 \,. \tag{3.11}$$

Moreover, the chosen network solution remains stable on the three-dimensional hypersurface if the determinant of the metric tensor for the corresponding hypersurface satisfies

$$g_3 = g_{11}(g_{22}g_{33} - g_{23}^2) - g_{12}(g_{12}g_{33} - g_{13}g_{23}) + g_{13}(g_{12}g_{23} - g_{13}g_{22}) > 0. \tag{3.12}$$

In this viewpoint, an arbitrary network system turns out to be stable as a hypersurface of dimension $m \leq 2n$ if the relevant principal minors and the determinant of the metric tensor remain positive-definite quantities. We thus see that the full configuration will be stable against simultaneous fluctuations of the network parameters if the underlying determinant of the metric tensor remains a positive-definite quantity over the range of interest of the considered parameters.

In the above definition, a network configuration is said to be completely stable if the set

$$\mathscr{B} := \{g_{ij}, g_i, g; x_i \in M_{2n}, \ i = 1, 2, \ldots, 2n\} \tag{3.13}$$

remains positive definite. It is worth mentioning that a network is stabilized if an arbitrary parameter $x_i \in M_{2n}$ also satisfies the same requirement, i.e., if it is an element of the set $\mathscr{B}$.

The parameters in a set $\{x_i \in M_{2n}\}$ are said to be (hyper)correlated if the components of the underlying Riemann covariant tensor R_{ijkl} are non-vanishing for a given set of indices i, j, k, l. In particular, the scalar curvature gives an average correlation volume for the constituent configuration of network parameters. The present research aims to forecast the values of network configurations associated with the power flow. Subsequently, we shall focus on the (in)stability domain of the relevant parameters.

From the power network perspective, there are no critical points in the parameter space in the set

$$C_p := \{x_1^0, x_2^0 \in \mathscr{C}\} \tag{3.14}$$

such that the corresponding power $P(x_1^0, x_2^0) = 0$. In this case, if there are some limiting values of the parameters for which the network burns, the network configuration thus obtained becomes an unstable system. A similar analysis is also straightforward for multi-component complex network configurations. In particular, we note that there does not exist a set of fixed parameters $\{x_1^0, x_2^0, \ldots, x_n^0\}$ such that the $P(x_i^0) = 0$. Consequently, the daughter system yields an unstable network configuration. Notice further that the present analysis is not limited to single-component networks. However, the investigation in question can indeed be carried out for arbitrary daughter networks. This offers the novel prospect of finding stable power flow solutions in the theory of finite parameter multi-component networks.

Furthermore, an extension of the present investigation could be made for general finite network configurations by defining the effective Hessian $g_{ij} = D_i D_j P(x_i)$ of the net power. The analogy follows from intrinsic manifolds and moduli space geometry in string theory [51–56]. Although the underlying physical interpretations of network theory may not remain quite the same as those of the chosen component network configurations, it should be noted that the parameters defining the underlying parameter manifold may not be globally stabilized in general. It would thus be natural to extend an understanding of the power flows, something we have thus studied, in the context of the simplest network models, defined by the leading order power network configurations. Nevertheless, the leading order power flow configuration is particularly well-suited for the present analysis, defining an intrinsic quadratic form which assumes no particular assignment of the individual components of the chosen network. The present analysis therefore considerably simplifies the underlying geometric computations and the possible investigation of the transformations concerning the power flow parameter.

At this juncture, it is worth noting that our geometrical analysis may equally be applied to other solutions of the network flow problem which possess a definite multi-component network configuration. Interestingly, the various extensions of the intrinsic geometry may further be analyzed apart from the multi-component networks. In fact, these investigations yield a set of realizations for the possible values of the parameters of the underlying multi-component network configurations that are intriguing from the standpoint of reliability and voltage (in)stability phenomena. Apart from the implications following from the leading order power flow, there exists a wide class of network theories, with and without the ripple factor, current filtering, and heating effects, as well as with single- or multi-component configurations, that might be further explored under the agenda of the present investigation.

The intrinsic space of parameters has provided crucial insights into the geometric understanding of a class of network configurations. Thus, it could be instructive to pursue geometric and algebraic approaches associated with various types of power flow, with the inclusion of arbitrary finite components, as well as different parameter dimensions. In particular, the underlying parameter space analysis can be investigated for the attractor (in)stability in a critical situation when the chosen network is near the black-out state. Importantly, the network configuration case may be investigated under parametric transformations involving certain elementary invariant parameters attached to a network system. Thus, one may intrinsically analyze the black-out phenomenon from the perspective of intrinsic/extrinsic geometry.

From the perspective of finite component complex networks, with parameters evaluated in the complex field, a finite collection of open sets can be constructed at a given point in the space of parameters. We then get the underlying manifold coordinates in the network configurations. For such a network configuration, we wish to analyze domains of (in)stability by focusing our attention on the role of the intrinsic Riemannian geometry. In order to keep arbitrary electric parameters, we shall consider the network configuration in its most general form. It is worth recalling that the limiting stability of the underlying fluctuations, causing some disturbances, requires a given sign of the relevant electric parameters. Consequently, we shall first illustrate the stabilization properties of network configurations with fewer constituent parameters. In general, we emphasize that the main purpose of the present research is to illustrate the phenomenological properties of such a network stabilization. In the sequel to this examination, we first focus our attention on the two- and three-parameter network configurations. However, one may indeed extend the present considerations to arbitrary parameter configurations. For the case of networks with two real parameters, this follows by considering the Hessian of the net power flow.

What follows next is that the geometric nature of the network system may be characterized by the parameter carried by the chosen configuration. Thus, depending on the sign of the underlying parameters, we shall analyze the intrinsic geometric issues and thereby describe the nature of the black-out phenomenon for network configurations with two and three real parameters. In the following chapters, we discuss in particular network reliability and voltage stability, along with possible insights arising from the use of real intrinsic geometry.

After this brief introduction, we shall now give a systematic analysis of the underlying stability structures of the parametric fluctuations of complex networks whose parameter space can be characterized as a real manifold. In particular, the state-space formulation of the network fluctuation problem will enable us to stabilize the fluctuations in the system through an effective power across a network component. In a practical power system, the power flow occurs at some dropping voltage across a component, and hence our proposal ensures the limiting instability. For a given effective power of the network, analysis of the fluctuations is well-suited for determining the allowed flow regimes of the parameters. For a given network configuration, we shall illustrate in Chap. 8 that the parametric fluctuations about a steady-state regime are governed by power-sum-type flow (in)stabilities. The parametric correction properties offer an understanding of which domains can be considered as stable domains of the network system.

References

1. R. Billinton, R.J. Ringle, A.J. Wood, *Power-System Reliability Calculations* (MIT Press Classics Series, September, 1973)
2. A. Chakraborty, P. Sen, An analytical investigation of voltage stability of an EHV transmission network based on load flow analysis. J. Inst. Eng. (India) Electr. Eng. Div. **76**, (1995)
3. P. Kundur, *Power System Stability and Control*, EPRI Power System Engineering Series. (McGraw-Hill, New York, 1994), p. 328
4. H. Frank, B. Landstorm, Power factor correction with thyristor-controlled capacitors. ASEA J. **45**, 180–184 (1971)
5. R. Rajarman, F. Alvarado, A. Maniaci, R. Camfield, S. Jalali, Determination of location and amount of series compensation to increase power transfer capability. IEEE Trans. Power Syst. **13**(2), 294–300 (1998)
6. H. Almasoud, Shunt capacitance for a practical 380 kV system. Int. J. Electr. Comput. Sci. (IJECS) **9**(10), 23–27 (2009)
7. G. Ruppeiner, Riemannian geometry in thermodynamic fluctuation theory. Rev. Mod. Phys. **67**, 605 (1995); [Erratum **68**, 313 (1996)]
8. G. Ruppeiner, Thermodynamic curvature and phase transitions in Kerr-Newman black holes. Phy. Rev. D **78**, 024016 (2008)
9. G. Ruppeiner, Thermodynamics: a Riemannian geometric model. Phys. Rev. A **20**, 1608 (1979)
10. G. Ruppeiner, Thermodynamic critical fluctuation theory? Phys. Rev. Lett. **50**, 287 (1983)
11. G. Ruppeiner, New thermodynamic fluctuation theory using path integrals. Phys. Rev. A **27**, 1116 (1983)
12. G. Ruppeiner, C. Davis, Thermodynamic curvature of the multicomponent ideal gas. Phys. Rev. A **41**, 2200 (1990)
13. T. Sarkar, G. Sengupta, B.N. Tiwari, On the thermodynamic geometry of BTZ black holes. JHEP **0611**, 015 (2006); arXiv:hep-th/0606084v1
14. T. Sarkar, G. Sengupta, B.N. Tiwari, Thermodynamic geometry and extremal black holes in string theory. JHEP **0810**, 076 (2008); arXiv:0806.3513v1 [hep-th]

15. B.N. Tiwari, *On Generalized Uncertainty Principle* (LAP Academic Publication, Germany, 2011); ISBN 978-3-8465-1532-7; arXiv:0801.3402v2 [hep-th]
16. B.N. Tiwari, *Sur les corrections de la géométrie thermodynamique destrous noirs* (Éditions Universitaires Européennes, Germany, 2011); ISBN 978-613-1-53539-0; arXiv:0801.4087v2 [hep-th]
17. B.N. Tiwari, *Geometric Perspective of Entropy Function: Embedding, Spectrum and Convexity* (LAP Academic Publication, Germany, 2011); ISBN 978-3-8454-3178-9; arXiv:1108.4654v2 [hep-th]
18. S. Bellucci, B.N. Tiwari, On the microscopic perspective of black branes thermodynamic geometry. Entropy **12**, 2096 (2010); arXiv:0808.3921v1
19. S. Bellucci, B.N. Tiwari, An exact fluctuating 1/2 BPS configuration. Springer J. High Energy Phys. **05**, 23 (2010); arXiv:0910.5314v1
20. S. Bellucci, B.N. Tiwari, State-space correlations and stabilities. Phys. Rev. D **82**, 084008 (2010); arXiv:0910.5309v1 [hep-th]
21. S. Bellucci, B.N. Tiwari, Thermodynamic geometry and Hawking radiation. JHEP **30**, 1011 (2010); arXiv:1009.0633v1 [hep-th]
22. S. Bellucci, B.N. Tiwari, Black strings, black rings and state-space manifold. Int. J. Mod. Phys. A **26**(32), 5403–5464 (2011); arXiv:1010.3832v2 [hep-th]
23. S. Bellucci, B.N. Tiwari, State-space manifold and rotating black holes. JHEP **118**, 1011 (2011); arXiv:1010.1427v1 [hep-th]
24. J.E. Aman, I. Bengtsson, N. Pidokrajt, Flat information geometries in black hole thermodynamics. Gen. Rel. Grav. **38**, 1305–1315 (2006); arXiv:gr-qc/0601119v1
25. J. Shen, R.G. Cai, B. Wang, R.K. Su, Thermodynamic geometry and critical behavior of black holes. Int. J. Mod. Phys. A **22**, 11–27 (2007); arXiv:gr-qc/0512035v1
26. J.E. Aman, I. Bengtsson, N. Pidokrajt, Geometry of black hole thermodynamics. Gen. Rel. Grav. **35**,1733 (2003); arXiv:gr-qc/0304015v1
27. J.E. Aman, N. Pidokrajt, Geometry of higher-dimensional black hole thermodynamics. Phys. Rev. D **73**, 024017 (2006); arXiv:hep-th/0510139v3
28. F. Weinhold, Metric geometry of equilibrium thermodynamics. J. Chem. Phys. **63**, 2479 (1975); doi:10.1063/1.431689
29. F. Weinhold, Metric geometry of equilibrium thermodynamics. II. Scaling, homogeneity, and generalized Gibbs-Duhem relations. J. Chem. Phys. **63**, 2482 (1975)
30. S. Bellucci, V. Chandra, B.N. Tiwari, On the thermodynamic geometry of hot QCD. Int. J. Mod. Phys. A **26**, 43–70 (2011); arXiv:0812.3792v1 [hep-th]
31. S. Bellucci, V. Chandra, B.N. Tiwari, A geometric approach to correlations and quark number susceptibilities; Mod. Phys. Lett. A **27**, 1250055 (2012) [6 pages] DOI: 10.1142/S0217732312500551, arXiv:1010.4405v1 [hep-th]
32. S. Bellucci, V. Chandra, B.N. Tiwari, Thermodynamic geometric stability of quarkonia states. Int. J. Mod. Phys. A **26**, 2665–2724 (2011); arXiv:1010.4225v2 [hep-th]
33. A. Strominger, C. Vafa, Microscopic origin of the Bekenstein-Hawking entropy. Phys. Lett. B **379**, 99–104 (1996); arXiv:hep-th/9601029v2
34. A. Sen, Extremal black holes and elementary string states. Mod. Phys. Lett. A **10**, 2081–2094 (1995); arXiv:hep-th/9504147v2
35. A. Dabholkar, Exact counting of black hole microstates. Phys. Rev. Lett. **94**, 241–301 (2005); arXiv:hep-th/0409148v2
36. L. Andrianopoli, R. D'Auria, S. Ferrara, Flat symplectic bundles of N-extended supergravities, central charges and black-hole, entropy; CERN Preprint CERN-TH/97–180, unpublished, arXiv:hep-th/9707203v1
37. A. Dabholkar, F. Denef, G.W. Moore, B. Pioline, Precision counting of small black holes. JHEP **0510**, 096 (2005); arXiv:hep-th/0507014v1
38. A. Dabholkar, F. Denef, G.W. Moore, B. Pioline, Exact and asymptotic degeneracies of small black holes. JHEP **0508**, 021 (2005); arXiv:hep-th/0502157v4
39. A. Sen, Stretching the horizon of a higher dimensional small black hole. JHEP **0507**, 073 (2005); arXiv:hep-th/0505122v2

40. J.P. Gauntlett, J.B. Gutowski, C.M. Hull, S. Pakis, H.S. Reall, All supersymmetric solutions of minimal supergravity in five dimensions. Class. Quant. Grav. **20**, 4587–4634 (2003); arXiv:hep-th/0209114v3

41. J.B. Gutowski, H.S. Reall, General supersymmetric AdS5 black holes. JHEP **0404P**, 048 (2004); arXiv:hep-th/0401129v3

42. I. Bena, N.P. Warner, One ring to rule them all ... and in the darkness bind them?. Adv. Theor. Math. Phys. **9P**, 667–701 (2005); arXiv:hep-th/0408106v2

43. J.P. Gauntlett, J.B. Gutowski, General concentric black rings. Phys. Rev. D **71**, 045002 (2005); arXiv:hep-th/0408122v3

44. S. Ferrara, R. Kallosh, A. Strominger, $N = 2$ extremal black holes. Phys. Rev. D **52**, R5412–R5416 (1995); arXiv:hep-th/9508072v3

45. A. Strominger, Macroscopic entropy of $N = 2$ extremal black holes. Phys. Lett. B **383**, 39–43 (1996); arXiv:hep-th/9602111v3

46. S. Ferrara, R. Kallosh, Supersymmetry and attractors. Phys. Rev. D **54**, 1514–1524 (1996); arXiv:hep-th/9602136

47. S. Ferrara, G.W. Gibbons, R. Kallosh, Black holes and critical points in moduli space. Nucl. Phys. B **500**, 75–93 (1997); arXiv:hep-th/9702103

48. S. Bellucci, S. Ferrara, A. Marrani, Attractors in black, Fortsch. Phys. **56**, 761 (2008); arXiv:0805.1310

49. G. Radman, R.S. Raje, Power flow model/calculation for power system with multiple FACTS controllers. Elsevier Sci. Dir. Electr. Power Syst. Res. **77**, 1521–1531 (2007)

50. J. Grainger Jr., W. Stevenson, *Power System Analysis*, 1st edn. (McGraw-Hill Science, Engineering, Math, New York, 1994)

51. E. Calabi, A construction of non-homogeneous Einstein metrics. Proc. Symp. Pure Math. (AMS, Providence) **27**, 17–24 (1975)

52. J. Li, S.T. Yau, *Hermitian Yang-Mills Connections on Non-Kähler Manifolds*, Mathematical Aspects of String Theory (World Scientific, Singapore, 1987)

53. A. Klemm, S. Theisen, Considerations of one-modulus Calabi-Yau compactifications: Picard-Fuchs equations, Kähler potentials and mirror maps. Nucl. Phys. B **389**, 153–180 (1983); arXiv:hep-th/9205041v1

54. P.S. Aspinwall, The Landau-Ginzburg to Calabi-Yau dictionary for D-branes. J. Math. Phys. **48**, 082304 (2007); arXiv:hep-th/0610209v2

55. A. Ricco, Brane superpotential and local Calabi-Yau manifolds. Int. J. Mod. Phys. A **23**, 2187–2189 (2008); arXiv:0805.2738v1 [hep-th]

56. G. Gibbons, R. Kallosh, B. Kol, Moduli, scalar charges, and the first law of black hole thermodynamics. Phys. Rev. Lett. **77**, 4992–4995 (1996); arXiv:hep-th/9607108v2

Chapter 4
A Test of Network Reliability

In this chapter, we focus on the aforementioned techniques of network theory and real intrinsic geometry. In order to carry out this investigation, we first formulate the problem as in the previous chapter. We then describe the details of this innovation for two-parameter single-component networks and show in the sequel that the method works in general. In particular, we evaluate the network reliability and test the results in order to demonstrate the accuracy of the proposed solution.

Regarding the proposed geometric state-space model, we note that nearest neighbor pure pair correlations decay as the inverse square of the chosen components, while mixed pair correlations decay only as the inverse of the chosen components [1–6]. The specific motivation in the present case is to investigate electrical configurations, where one wants to study the issue of network reliability and voltage stability of the circuit. The limiting network reliability of an arbitrary transmission line and limiting voltage stability of an arbitrary finite number of components are effectuated with a fixed set of resistances, inductances, and reactive capacitances.

The first step is to examine the intrinsic Weinhold geometry of the fluctuations about an equilibrium fixed point (X_L, r), for the chosen LR component of specific transmission lines connected to a disturbed bus.

As a function of X_L and r, the Weinhold metric tensor of the reactive space, defined by the maxima of the effective power, is given by

$$g_{ij} = \begin{pmatrix} \frac{\partial^2 S}{\partial X_L^2} & \frac{\partial^2 S}{\partial X_L \partial r} \\ \frac{\partial^2 S}{\partial X_L \partial r} & \frac{\partial^2 S}{\partial r^2} \end{pmatrix}. \tag{4.1}$$

The Riemann–Christoffel connection components are defined by

$$\Gamma_{ijk} = g_{ij,k} + g_{ik,j} - g_{jk,i}, \tag{4.2}$$

where $g_{ij,k}$ denotes the partial derivative of the local flow component g_{ij} with respect to the network parameter x_k. It follows that in the case of the RL circuit, there is only one nonzero component of the Riemann tensor, viz.,

S. Bellucci et al., *Geometrical Methods for Power Network Analysis*,
SpringerBriefs in Electrical and Computer Engineering,
DOI: 10.1007/978-3-642-33344-6_4, © The Author(s) 2013

$$R_{X_L r X_L r} = \frac{N}{D},$$ (4.3)

where

$$N = S_{rr}(S_{X_L rr}S_{X_L X_L X_L} - S_{rX_L X_L}) + S_{X_L X_L}(S_{rX_L X_L}S_{rrr} - S_{X_L rr})$$
$$+ S_{rX_L}(S_{rX_L X_L}S_{X_L rr} - S_{rrr}S_{X_L X_L X_L})$$ (4.4)

and

$$D = 4(S_{rr}S_{X_L X_L} - S_{rX_L}^2).$$ (4.5)

Here, the subscripts on the effective power S indicate the corresponding partial derivatives with respect to the network parameters.

For the above type of two-component network, the Ricci scalar is given by

$$R = \frac{2}{\det g}R_{X_L r X_L r}.$$ (4.6)

From the hypothesis of power system planning, the intrinsic geometry gives the following local and global stability criteria:

- A novel aspect of our proposal is that its methodology allows one to predict both the reliability and the stability of the desired power system and its operation.
- For an arbitrary component finite network, the proposed intrinsic geometric formulation shows that reliability corresponds to hyperspace positivity, while stability is determined by positivity of the determinant of the state-space metric tensor.
- The global network reliability and global voltage stability are determined from the regularity properties of the scalar curvature of the underlying reactive-space invariants.

From the perspective of network power flow, our analysis shows that it is legitimate to analyze the evolution of the reactive component, infinitesimally fluctuating about an equilibrium configuration. The state of the art is then generically defined as the sum of the infinitesimal evolutions of an ensemble of network components. This is equivalent to considering the nearest neighbor interactions of the chosen component(s). For the foregoing consideration, we exploit the fact that a chosen component is in an equilibrium configuration if the node voltages satisfy $|V_i| = 1$ for all values of i.

The network reliability can be tested using the intrinsic geometry setup. In order to illustrate with a real application of our intrinsic geometric proposal, let us consider the case of the RL component. Without loss of generality, we can choose δ to be a constant at equilibrium. Note that the constant phases can be adjusted as required. This provides freedom to choose to work either with the cosine or the sine of the phase. Accordingly, (2.1) shows that the power liberated by the ith line is

$$P_i = Y_{i1}V_1V_i \cos\left(\delta_i - \delta_1 - \tan^{-1}\frac{X_{L_{i1}}}{r_{i1}}\right).$$ (4.7)

Using the standard trigonometric identity

$$\cos\left(\tan^{-1}\frac{X}{r}\right) = \frac{r}{\sqrt{r^2 + X^2}}, \tag{4.8}$$

and the fact that $|V_i| = 1$ at equilibrium, and defining the impedance of the component by

$$|Y| = |Z|^{-1}, \tag{4.9}$$

we find that the power liberated by an arbitrary LR component has the simple expression

$$P(r, L) = \frac{r}{r^2 + \omega^2 L^2}. \tag{4.10}$$

Thus, the components of the metric tensor on the inductive manifold are

$$\begin{aligned}
g_{rr} &= 2r\frac{r^2 - 3\omega^2 L^2}{(r^2 + \omega^2 L^2)^3}, \\
g_{rL} &= 2\omega^2 L\frac{3r^2 - \omega^2 L^2}{(r^2 + \omega^2 L^2)^3}, \\
g_{LL} &= -2r\omega^2 L\frac{r^2 - 3\omega^2 L^2}{(r^2 + \omega^2 L^2)^3}.
\end{aligned} \tag{4.11}$$

We observe that fluctuations in the resistance and inductance have competing effects on the reliability. For the typical transmission line $r \ll X_L$, because a high value of r leads to a high loss of real power. Further losses due to the inductive component can be controlled by appropriate compensation. Thus, from the fact that the fluctuation in r (due to the uncertain real load) beyond a specified value of the inductance makes the inductive configuration unreliable.

This interesting intrinsic picture goes further. We find that the determinant of the metric tensor is given by

$$\det g = -4\frac{\omega^2}{(r^2 + \omega^2 L^2)^3}. \tag{4.12}$$

More explicitly, $g_{rr} > 0$ shows that a relatively large resistive load is reliable, whereas the condition $g_{LL} < 0$ implies that the inductive load is reliable, and this is affected by the resistive loading of the system. The boundary of reliability for choosing parameters is thus given by the following $\{r, L\}$ curve:

$$r^2 - 3\omega^2 L^2 = 0. \tag{4.13}$$

Appropriate reactors can therefore be designed for varying resistive demand and also to reduce the resistance of the line, as required for parallel lines, in such a way as to keep the power flow of the network reliable.

Negativity of the determinant of the metric tensor det g shows that the entire power system becomes unreliable, if both the resistive and the inductive components, viz., $\{r, L\}$, vary simultaneously. Thus, to further reduce the inductance effect of a transmission line or lines connected to a bus bar, appropriate compensation is required. This will be computed in the next chapter.

From (3.8), we see that the Ricci scalar curvature vanishes identically and we have

$$R\,(r, L) = 0, \quad \forall\,(r, L) \,\in\, \mathcal{M}_2, \tag{4.14}$$

for the RL metric tensor as defined in (4.11). The vanishing of the intrinsic scalar curvature demonstrates that, as a finite union of LR components, the network is free from global instabilities, so the intrinsic geometric analysis shows that an ensemble of arbitrary LR components corresponds to a globally reliable power system.

References

1. H. Frank, B. Landstorm, Power factor correction with thyristor-controlled capacitors. ASEA J. **45**, 180–184 (1971)
2. P.M. Anderson, R.G. Farmer, *Series Compensation of Power Systems* (Fred Laughter and PBLSH Inc., Encinitas, 1996) ISBN-10: 1888747013
3. J. Stones, A. Collinson, Introduction to power quality. Power Eng. J. **15**(2), 58–64 (2001)
4. T.J. Miller, *Reactive Power Control in Electrical Systems* (Wiley, New York, 1982)
5. G. Radman, R.S. Raje, Power flow model/calculation for power system with multiple FACTS controllers. Electr. Power Syst. Res. (Elsevier, ScienceDirect) **77**, 1521–1531 (2007)
6. T.J.E. Miller (ed.), *Reactance Power Control in Electric Systems* (Wiley, New York, 1982)

Chapter 5
A Test of Voltage Stability

In this chapter, we extend the techniques of intrinsic geometry and examine the problem of voltage stability in network theory. In order to carry out this investigation, we determine stability domains by first specifying the three-parameter model for the reformulation of the problem. We then consider single-component LCR networks, showing how voltage stability is achieved according to the results of the present model. From the outset of the present investigation, we offer specific remarks and outlook for future research.

The metric tensor on the reactive state-space manifold is defined by

$$g_{ij} = \partial_i \partial_j S(x_1, x_2, \ldots, x_n), \quad x_1, x_2, \ldots, x_n \in M_n. \tag{5.1}$$

This can be used to characterize the LCR fluctuations. The above metric tensor may be further supported by the standard Taylor expansion of the effective (complex) power liberated in the network.

From the above considerations, it follows that the power flow of a general network reduces to the power fluctuations on the reactive configuration, involving the resistances and reactance in the case of network reliability, or both the inductance and capacitances in the case of the voltage stability. In both the above cases, the parametric fluctuations yield the standard basis for the associated intrinsic reactive Riemannian manifold M_n.

As a non-linear combination of power fluctuations, the scalar curvature allows one to set up global notions of network reliability and voltage stability for the LR and LCR power systems.

In fact, it turns out that the scalar curvature $R(\{x_i\})$ scales as

$$R \sim \zeta^d(\{x_i\}), \tag{5.2}$$

where d is the spatial dimension of the underlying system and $\zeta(\{x_i\})$ fixes the correlation scale for the reliability and voltage stability of the chosen LR or LCR component of the network. Divergence of the invariant scalar curvature indicates possible critical behavior or phase transitions in the underlying system, and at such

S. Bellucci et al., *Geometrical Methods for Power Network Analysis*,
SpringerBriefs in Electrical and Computer Engineering,
DOI: 10.1007/978-3-642-33344-6_5, © The Author(s) 2013

a critical point, the configuration is prone to have a breakdown for the power flow scenarios considered here.

The local and global network reliability and voltage stability of the electrical component(s) are thus taken into account in a unified manner with the inclusion of non-linear effects of the network fluctuations. Such variations can in principle be the ripple factor, current filtering, and heating effects, which are simultaneously taken into account in the present analysis. In contrast to linear transient analysis, the present model gives a fast prediction of the integrated behavior of the desired network component(s).

With the notions introduced in the previous chapter for the network reliability, the tuning of the line parameter carried out by r and L may similarly be extended to the LCR components by adding a nonzero capacitance to the LR component. This yields a self-consistent method for maintaining a constant voltage profile. In this case, from the power flow equations, we find that the real and reactive power liberated on the ith bus through the line $i = 1$ is

$$
\begin{aligned}
P_i &= Y_{i1} V_1 V_i \cos\left(\delta_i - \delta_1 - \tan^{-1} \frac{X_{L_i 1} - X_{C_i 1}}{r_{i1}}\right), \\
Q_i &= Y_{i1} V_1 V_i \sin\left(\delta_i - \delta_1 - \tan^{-1} \frac{X_{L_i 1} - X_{C_i 1}}{r_{i1}}\right).
\end{aligned}
\tag{5.3}
$$

At local equilibrium, an appropriate choice of the load angles is $\delta_i = \delta_1$. Thus, the unified power liberated on the ith bus due to the connected lines reduces to

$$
S(r, X_L, X_C) = \frac{r + X_L - X_C}{r^2 + (X_L - X_C)^2}.
\tag{5.4}
$$

From the above equation, direct substitution of X_L and X_C shows that the total effective power through the LCR component takes the form

$$
S(r, L, C) = \frac{r + \omega L - 1/\omega C}{r^2 + (\omega L - 1/\omega C)^2}.
\tag{5.5}
$$

As mentioned for the LR component, the components of the metric tensor can be computed from the definition of the Hessian matrix $\mathrm{Hess}\big(P(r, L, C)\big)$. Thus, we see that the components of the metric tensor, combining the real and imaginary power flow fluctuations, are given by

$$
\begin{aligned}
g_{rr} &= 2\omega^3 C^3 \frac{\tilde{g}_{rr}}{(r^2 \omega^2 C^2 + \omega^4 L^2 C^2 - 2\omega^2 LC + 1)^3}, \\
g_{rL} &= -2\omega^4 C^3 \frac{\tilde{g}_{rL}}{(r^2 \omega^2 C^2 + \omega^4 L^2 C^2 - 2\omega^2 LC + 1)^3}, \\
g_{rC} &= -2\omega^2 C \frac{\tilde{g}_{rC}}{(r^2 \omega^2 C^2 + \omega^4 L^2 C^2 - 2\omega^2 LC + 1)^3},
\end{aligned}
\tag{5.6}
$$

$$g_{LL} = -2\omega^3 C \frac{\tilde{g}_{LL}}{(r^2\omega^2 C^2 + \omega^4 L^2 C^2 - 2\omega^2 LC + 1)^3},$$

$$g_{LC} = -2\omega^3 C \frac{\tilde{g}_{LC}}{(r^2\omega^2 C^2 + \omega^4 L^2 C^2 - 2\omega^2 LC + 1)^3}, \qquad (5.7)$$

$$g_{CC} = -2\omega^2 \frac{\tilde{g}_{CC}}{(r^2\omega^2 C^2 + \omega^4 L^2 C^2 - 2\omega^2 LC + 1)^3}.$$

Here, the numerators of the components of the metric tensor reduce to

$$\tilde{g}_{rr} = r^3\omega^3 C^3 - 3r\omega^5 C^3 L^2 + 6r\omega^3 C^2 L - 3r\omega C + 3r^2\omega^4 C^3 L - 3r^2\omega^2 C^2$$
$$- \omega^6 L^3 C^3 + 3\omega^4 L^2 C^2 - 3\omega^2 LC + 1, \qquad (5.8)$$

$$\tilde{g}_{rL} = r^3\omega^3 C^3 - 3r\omega^5 C^3 L^2 + 6r\omega^3 C^2 L - 3r\omega C - 3r^2\omega^4 C^3 L + 3r^2\omega^2 C^2$$
$$+ \omega^6 L^3 C^3 - 3\omega^4 L^2 C^2 + 3\omega^2 LC - 1, \qquad (5.9)$$

$$\tilde{g}_{rC} = r^3\omega^3 C^3 - 3r\omega^5 C^3 L^2 + 6r\omega^3 C^2 L - 3r\omega C - 3r^2\omega^4 C^3 L + 3r^2\omega^2 C^2$$
$$+ \omega^6 L^3 C^3 - 3\omega^4 L^2 C^2 + 3\omega^2 LC - 1, \qquad (5.10)$$

$$\tilde{g}_{LL} = r^3\omega^3 C^3 - 3r\omega^5 C^3 L^2 + 6r\omega^3 C^2 L - 3r\omega C + 3r^2\omega^4 C^3 L - 3r^2\omega^2 C^2$$
$$- \omega^6 L^3 C^3 + 3\omega^4 L^2 C^2 - 3\omega^2 LC + 1, \qquad (5.11)$$

$$\tilde{g}_{LC} = r^3\omega^3 C^3 - 3r\omega^5 C^3 L^2 + 6r\omega^3 C^2 L - 3r\omega C - 3r^2\omega^4 C^3 L + 3r^2\omega^2 C^2$$
$$+ \omega^6 L^3 C^3 - 3\omega^4 L^2 C^2 + 3\omega^2 LC - 1, \qquad (5.12)$$

$$\tilde{g}_{CC} = \omega L - r - 3r^4\omega C + 3\omega^5 C^2 L^3 + 3r^3\omega^2 C^2 + 3r^2\omega^3 C^2 L + 3r\omega^4 C^2 L^2$$
$$- 2r\omega^6 L^3 C^3 - 2r^3\omega^4 C^3 L - 3\omega^3 L^2 C - \omega^7 C^3 L^4 + r^4\omega^3 C^3. \qquad (5.13)$$

For the voltage stability of the ith bus, the issues in the following sections are important as regards the intrinsic geometry.

5.1 Surface Stability

The principal minor P_2, defining the stability of LC surface variations, is depicted in Fig. 5.1 as a function of the inductance L and capacitance C for $r = 0$. This shows that an appropriate choice of C is required for relatively small values of L. For instance, C should be smaller than 0.5 p.u. Furthermore, it turns out that the deviation of the complex power grows rapidly, i.e., as e^{22}, to a large negative value. The general expression for the minor is not very elegant. Specifically, for $r = 0$, we find that the surface minor of the LC oscillations becomes unstable with the following value for the limiting minor:

Fig. 5.1 Surface minor plotted as a function of the power
factors L and C, to test the
voltage stability of power fluctuations in electrical networks

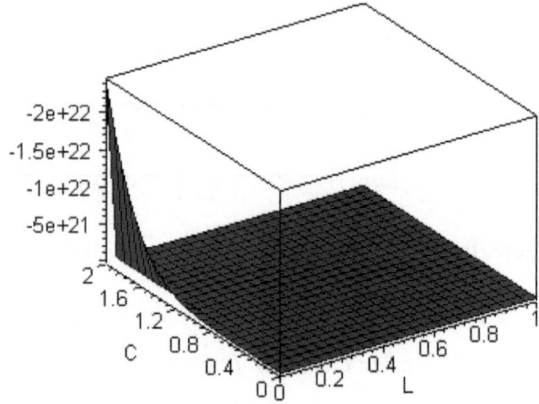

$$P_2 = -4\omega^6 C^4 \frac{\tilde{P}_2}{(1 + \omega^4 L^2 C^2 - 2\omega^2 LC)^6},\tag{5.14}$$

where

$$\tilde{P}_2 = 1 - 20\omega^6 L^3 C^3 - 20\omega^8 L^3 C^5 + 15\omega^{10} L^4 C^6$$
$$+ 15\omega^8 L^4 C^4 - 6\omega^4 C^3 L - 6\omega^{10} L^5 C^5 - 6\omega^2 LC$$
$$- 6\omega 12 L^5 C^7 + 15\omega^4 L^2 C^2 + 15\omega^6 L^2 C^4$$
$$+ \omega^{14} L^6 C^8 + \omega^{12} L^6 C^6 + \omega^2 C^2.\tag{5.15}$$

5.2 Volume Stability

The determinant of the metric tensor defining voltage stability of the LRC variations
has been depicted in Fig. 5.2. The determinant of the metric tensor is plotted as a
function of the inductance L and capacitance C for $r = 0$. It shows that, for relatively
small values of L, the corresponding value of C should be less than 0.5 p.u. Both
the surface and volume stability of the LCR oscillations thus remain the same for
the zero value of the resistance. Over a range of values of $\{L, C, r = 0\}$, it is worth
mentioning that the qualitative features of the determinant of the metric tensor remain
the same as for the surface minor. In the limit of $L = 0$ and $r = 0$, we notice that
the determinant of the metric tensor reduces to the expression

$$\det g = 8(1 - 3\omega^2 C^2)\omega^7 C^3.\tag{5.16}$$

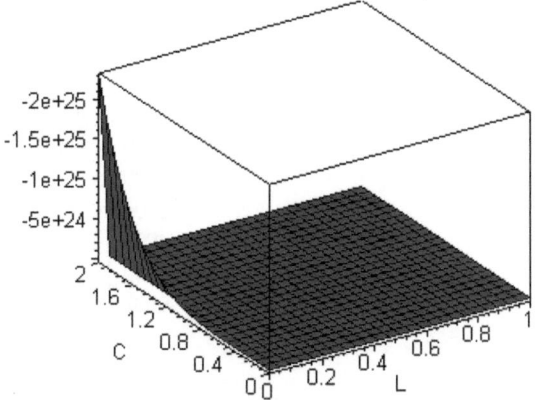

Fig. 5.2 Determinant of the metric tensor plotted as a function of the power factors L and C, to test the voltage stability of power fluctuations in electrical networks

As we move to higher values of C as compared to L, we observe that the complex power fluctuations grow rapidly and reach a very high value. In contrast to what happens for network reliability, the deviation in the voltage stability grows to the order of e^{25}. For an AC frequency of $f = 50$ Hz, we notice further that the limiting $L = 0$, $r = 0$ determinant of the metric tensor remains positive for large C. Specifically, our analysis can provide highly stable choices for the values of L and C. Such a characterization allows for the required regulation of the power supply.

5.3 Global Stability

The state-space global stability of LCR network fluctuations is depicted in Fig. 5.3, as a function of the inductance L and capacitance C for the choice $r = 0$. It shows that the global nature of limiting LCR oscillations is acceptable for a certain range of the capacitance. In the limit of zero capacitance, the system shows a large global correlation of order e^3. For small L, e.g., $L < 0.4$, we find that such an LCR component is globally weakly correlated. However, for a relatively high value of L, the component demands a higher value of C.

As we move to higher values of C, we observe that the fluctuation of the total power ceases rapidly and the whole component becomes globally uncorrelated. The general expression for the curvature scalar is rather intricate. Interestingly, for the choice $L = 0$ and $r = 0$, we notice that the limiting curvature scalar takes the following form

$$R(L = 0, r = 0, C) = \frac{1}{2\omega C} \frac{-6 + 25\omega^2 C^2 - 51\omega^4 C^4}{1 - 3\omega^2 C^2}. \tag{5.17}$$

It is worth mentioning that the limiting system becomes unstable for the vanishing value of the determinant of the metric tensor. From (5.17), we see that the scalar

Fig. 5.3 The curvature scalar
plotted as a function of the
power factors L, C, to test the
voltage stability of power fluctuations in electrical networks

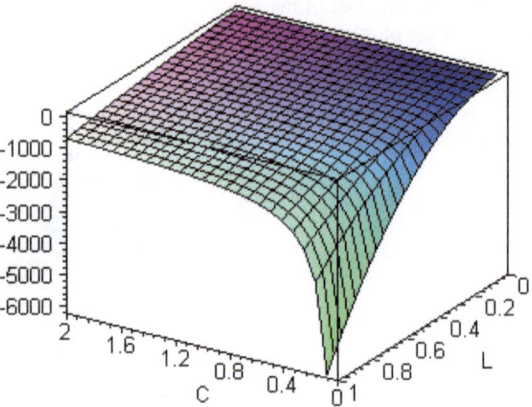

curvature of the limiting $L = 0$ and $r = 0$ configuration diverges, and thus we find
in this limit that the entire system becomes highly unstable. An identical situation
holds for general values of the network parameters L, C, and r.

Regarding Figs. 5.1, 5.2, and 5.3, where the third dimension shows the minor,
determinant of the metric tensor, and global scalar curvature of the system, we may
conclude for $r = 0$ that the most appropriate values of the capacitance should lie
in the range $0.1 < C < 0.5$. For the stability of the LCR component, this follows
from the fact that the surface minor and determinant of the metric tensor decrease
for $C > 0.5$, while for $0.1 < C$, the scalar curvature blows up. Thus, we can design
an appropriate compensator and reactor to maintain the system voltage.

5.4 Specific Remarks

5.4.1 Limiting Voltage Stability

The configuration with $r = 0$ corresponds to the limiting voltage stabilization, as
shown in the Figs. 5.1, 5.2, and 5.3. These diagrams describe the nature of the LC
fluctuations of the LCR component about a fixed AC base line, e.g., a constant
frequency of 50 or 60 Hz.

5.4.2 Limiting Reliability

The case $C \rightarrow 0$ corresponds to the limiting reliability of the LCR component. This
notion may be analyzed in a similar manner to the limit $r \rightarrow 0$. Such considerations
describe LR fluctuations of the LCR component about an almost fixed DC base. For

Table 5.1 Quantitative verification for the LR component, shown for the determinant of the metric tensor as a function of r and L, to test the voltage stability power fluctuations in electrical networks

T	r [p.u.]	L [p.u.]	det g
1	0.02	0.60	−852772.88
2	0.08	0.24	−208.19
3	0.06	0.18	−1169.78
4	0.06	0.68	−0.41
5	0.04	0.12	−13324.57
6	0.01	0.03	−54577464.92
7	0.08	0.024	−22 377 541.16

T stands for the transmission line

Table 5.2 Quantitative verification for the LR component shown for the surface minor, determinant of the metric tensor, and curvature scalar as a function of r, L, and C, to test the voltage stability of power fluctuations in electrical networks

T	r [p.u.]	L [p.u.]	C [p.u.]	P2	det g	R
1	0.02	0.60	0.30	−272.93	−11519.51	−4.92
2	0.08	0.24	0.025	−0.20e-2	0.68	36.89
3	0.06	0.18	0.020	−0.74e-3	0.27	46.58
4	0.06	0.68	0.020	−0.14e-2	0.79	41.45
5	0.04	0.12	0.015	−0.21e-3	0.96e-1	62.82
6	0.01	0.03	0.010	−0.38e-4	0.24e-1	95.40
7	0.08	0.024	0.025	−0.15e-2	0.39	39.90

T stands for the transmission line

a variable base inductance, the inductive fluctuations introduce a nonzero ripple in the LR component.

5.4.3 Non-linear Reliability and Stability

Regarding the validity of the parameterization of the LCR components, Fig. 5.3 shows that zero capacitance corresponds to a singular point of the reactive configuration. This accords with the regularity condition of the Jacobian matrix for the coordinate transformation of the fluctuating LCR component.

Here we find that the strongly correlated reactive space notions remain valid even for the limiting zero resistance component $r \to 0$. In this respect, Table 5.1 offers a quantitative verification for the LR component, while Table 5.2 gives a quantitative verification for the LCR component. Here, the symbol T stands for the transmission line.

To check the reliability of the transmission line, we propose an explicit test for the chosen component. From Table 5.1, we note that all the transmission lines have a negative determinant of the metric tensor. This indicates that there is a need to strengthen the transmission line by satisfying the $\{r, L\}$ curve condition

$$r^2 - 3\omega^2 L^2 = 0, \tag{5.18}$$

in order to obtain a reliable power supply. As mentioned in the last chapter, such a boundary of reliability allows one to choose the network parameters. For a given shunt capacitance, the corresponding stability test is depicted in Table 5.2. This shows that power flows in the transmission lines demanding certain variations make the system unstable. Compensation must therefore be provided for all the transmission lines except the first line. In this case, this illustrates the physical quantification of the fluctuations shown in the Figs. 5.1, 5.2, and 5.3.

Chapter 6
Phases of Power Network

In this chapter, we describe the intrinsic geometric design of power flow and the parametric stability of power networks by focusing our attention on the admissible values of the parameters $\{L, C, R\}$ for the real power flow, the imaginary power flow, and their arbitrary linear combinations as the unified description of the network power flow. In the context of the power flow equations [1, 2], the usefulness of the present investigation may appear to be somewhat limited. However, we shall show how our proposal could in principle be generically applied to all electrical networks, in order to achieve a better understanding of the phenomenon and importance of controlled power flow and network analysis. Intrinsic geometric considerations can provide strategic planning criteria for the effective use of power systems and voltage stability. We show that this notion follows from the standard laws of electrical circuits (for a review, see [2]). For an additional component, the criteria of voltage stability can then be used for optimal selection of the network parameters.

We thus formulate the problem and outline the notion of intrinsic Riemannian geometry. Investigation of the power flow equations relates the power flow to the network phases, and thereby yields the fixed point(s) for the stability analysis of the power network. Our intrinsic geometric model is designed to provide the critical values of the phases in an efficient way. It is worth mentioning that the present method is useful for determining the parameters, and thus the unstable modes of the network.

6.1 Network Power Flow

To optimize the power flow, we make the same assumptions as for the previous solutions and use the load flow equations [1, 2] in order to deal with the discussed issues of power flow fluctuations. The associated power conservation equations with real (resistive) and imaginary (reactive) branch parameters are

$$P_i = \sum |V_i||V_j||Y_{ij}|\Big[G_{ij}\cos(a_{ij} + \delta_j - \delta_i) + B_{ij}\sin(a_{ij} + \delta_j - \delta_i)\Big], \quad (6.1)$$

S. Bellucci et al., *Geometrical Methods for Power Network Analysis*,
SpringerBriefs in Electrical and Computer Engineering,
DOI: 10.1007/978-3-642-33344-6_6, © The Author(s) 2013

$$Q_i = \sum |V_i||V_j||Y_{ij}|\Big[G_{ij}\sin(a_{ij} + \delta_j - \delta_i) - B_{ij}\cos(a_{ij} + \delta_j - \delta_i)\Big]. \quad (6.2)$$

In these equations, the phases are defined by

$$\tan a_{ij} = \frac{X_{L_{ij}} - X_{C_{ij}}}{r_{ij}}. \quad (6.3)$$

The inverse set of impedances Z_{ij} and the voltage angles δ_j are

$$Y_{ij} = \frac{1}{r_{ij} + jX_{L_{ij}} - jX_{C_{ij}}}, \qquad \delta_j = \frac{V_j}{|V_j|}. \quad (6.4)$$

For the purposes of the following analysis, let us consider an arbitrary ith bus such that the steady-state condition is realized as $|V_i| = 1$, whence the underlying configuration reaches an equilibrium. The cases of present interest then reduce to standard network considerations. For this analysis, we consider that a lossless line is defined by $a_{ij} = \pm 90°$, where $+90°$ represents the ideal case. It then turns out that the network is purely inductive with $r = 0$. Notice that a realistic network would never reach the limit of zero resistance. In generic situations, the values of the a_{ij} vary from $+90°$ to $-90°$. Further, the phase $a_{ij} = -90°$ is not feasible in realistic situations, since the network possesses a finite capacitance. Thus, the phases for the inductor and capacitor circuits can be defined as

$$a_{(1)ij} = \tan^{-1}\frac{X_{L_{ij}}}{r_{ij}}, \quad a_{ij} = -90°,$$
$$a_{(2)ij} = \tan^{-1}\frac{X_{C_{ij}}}{r_{ij}}, \quad a_{ij} = 90°. \quad (6.5)$$

For general consideration of network fluctuations, we have nonzero values for the network parameters r, L, and C, and thus the general phase angle $a_{(3)ij}$ is defined as

$$a_{(3)ij} = \tan^{-1}\frac{X_{L_{ij}} - X_{C_{ij}}}{r_i j}. \quad (6.6)$$

We take into account the fact that the efficiency of the power flow on a transmission line is analyzed by the phases of the impedances pertaining to the transmission lines. Concerning the considerations of the next section, our method offers a non-linear characterization for the generic component of a realistic network, which we suppose to be neither a purely inductive nor a purely capacitive component.

6.2 Intrinsic Geometric Basis

In this section, we recall the motivation for intrinsic geometric analysis and set up the notations for the subsequent computations. Using the notation of the last section, a given network can reach a local equilibrium if we can fix one of the phases of the power network. The logic simply follows from the fact that the sum of the three angles of a trigone is a constant. To be specific, we illustrate the intrinsic geometry considerations for the case of two-parameter configurations.

To be concrete, let the parameters be $\{a_1, a_2\}$, and let $S(a_1, a_2)$ be a smooth function of the network (real, imaginary) phases as defined by (6.1) and (6.2) or any of their real combinations. For a given $S(a_1, a_2)$, the components of the correlation functions are described by the Hessian matrix $\mathrm{Hess}(S(a_1, a_2))$ of the generalized power function under the flow of the parameters. Given that the phases of the impedance pertaining to the transmission lines of a chosen network configuration are taken to be statistical, we see that the components of the intrinsic metric tensor are given by

$$g_{a_1a_1} = \frac{\partial^2 S}{\partial a_1^2}, \quad g_{a_1a_2} = \frac{\partial^2 S}{\partial a_1 \partial a_2}, \quad g_{a_2a_2} = \frac{\partial^2 S}{\partial a_2^2}. \tag{6.7}$$

These components are associated with the pair correlation functions of the relevant power flow. It is worth mentioning that the coordinates of the underlying power factor lie on the parameter surface, which in the statistical sense gives the origin of the fluctuations in the network. This is because the components of the metric tensor comprise the Gaussian fluctuations of the network power, which is itself a function of the parameters of the power configuration. For a given network, local stability of the underlying system requires both the principal components to be positive. In this respect, the diagonal components of the metric tensor, $\{g_{a_ia_i} | i \in 1, 2\}$ correspond to the heat capacities of the system, so they are required to remain positive-definite quantities:

$$g_{a_ia_i} > 0, \quad i = 1, 2. \tag{6.8}$$

From the perspective of intrinsic geometry, the stability properties of the network flows can thus be established from the positivity of the determinant of the metric tensor. For the Gaussian fluctuations of the two-charge equilibrium power configurations, the existence of a positive-definite volume form on the power surface imposes such a stability condition. Specifically, a power-supplying configuration is said to be stable if the determinant of the tensor, viz.,

$$\|g\| = S_{a_1a_1} S_{a_2a_2} - S_{a_1a_2}^2, \tag{6.9}$$

remains positive. For two-parameter networks, the geometric quantities corresponding to the chosen power elucidate the typical features of the Gaussian fluctuations about an ensemble of equilibrium states. As a global invariant, the intrinsic scalar curvature encodes the information included in the correlation volume of the underlying power fluctuations. Explicitly, the scalar curvature R takes the form

$$\begin{aligned} R = -\frac{1}{2(S_{a_1a_1} S_{a_2a_2} - S_{a_1a_2}^2)^2} \Big(&S_{a_2a_2} S_{a_1a_1a_1} S_{a_1a_2a_2} + S_{a_1a_2} S_{a_1a_1a_2} S_{a_1a_2a_2} \\ &+ S_{a_1a_1} S_{a_1a_1a_2} S_{a_2a_2a_2} + S_{a_1a_2} S_{a_1a_1a_1} S_{a_2a_2a_2} \\ &- S_{a_1a_1} S_{a_1a_2a_2}^2 - S_{a_2a_2} S_{a_1a_1a_2}^2 \Big). \end{aligned} \tag{6.10}$$

Notice that zero scalar curvature indicates that the power of the network fluctuates independently of the phases, while a divergent scalar curvature signifies a sort of phase transition, indicating an ensemble of highly correlated pixels of information on the power surface. In the case of black hole physics, Ruppeiner has interpreted the assumption that all the statistical degrees of freedom of a black hole live on the black hole event horizon as an indication that the state-space scalar curvature indicates the average number of correlated Planck areas on the event horizon of the black hole [3]. For the case of two-parameter systems, the above analysis of the surface shows that the scalar curvature and curvature tensor are related by

$$R(a_1, a_2) = \frac{2}{\|g\|} R_{a_1 a_2 a_1 a_2}. \tag{6.11}$$

The scalar curvature thus defined informs as to the nature of the long-range global correlation and underlying phase transitions originating from the power flow. In this sense, we anticipate that an ensemble of signals corresponding to the network are statistically interacting if the underlying power configuration has a nonzero scalar curvature. Incrementally, we may notice further that the configurations under present consideration are allowed to be effectively attractive or repulsive, and weakly interacting, in general. The intrinsic geometric analysis further provides a set of physical indications encoded in the geometrically invariant quantities, e.g., scalar curvature and other geometrically non-trivial objects. For the electrical network, the underlying analysis would involve an ensemble or subensemble of the equilibrium configuration forming a statistical basis about the Gaussian distribution. With this brief introduction, we shall now proceed to systematically analyze the underlying stability structures of the network fluctuations in the real and imaginary power flows on the network, together with their joint effects.

6.3 Real Power Flow

Let us first describe the intrinsic stability of the electrical network with a given power factor. By (6.1), the power defined with a set of desired corrections over the network power factors, chosen as the network variables a_1, a_2 for the present analysis, is given by

$$P(a_1, a_2) := \frac{V^2}{R_0 \big[1 + (\tan a_1 - \tan a_2)^2 \big]}. \tag{6.12}$$

The components of the correlation functions are described by the Hessian matrix $\text{Hess}\big(P(a_1, a_2)\big)$ of the relevant power under the tuning response function. By (6.12), the components of the metric tensor are

$$g_{a_1 a_1} = \frac{2c_2^3 V^2}{R_0} \frac{n_{11}^R}{r_{11}^R}, \quad g_{a_1 a_2} = -\frac{2c_2^2 c_1^2 V^2}{R_0} \frac{n_{12}^R}{r_{12}^R}, \quad g_{a_2 a_2} = \frac{2c_1^3 V^2}{R_0} \frac{n_{22}^R}{r_{22}^R}. \tag{6.13}$$

In this framework, we find that the geometric nature of the parametric (pair) correlations provides a way to formulate the (local) notion of fluctuating networks. Hence, the fluctuating parameters are easily ascertained in terms of the intrinsic parameters of the underlying network configurations. For a given network, it is evident that the principal components of the metric tensor signify self pair correlations, which are positive definite functions over a range of the parameters. In order to simplify the subsequent notation, let us define c_i and s_i by

$$c_i := \cos a_i , \quad s_i := \sin a_i , \quad i = 1, 2. \tag{6.14}$$

We thus arrive at the conclusion that the numerators of the local pair correlation functions are expressed by the following trigonometric polynomials:

$$
\begin{aligned}
n_{11}^R &:= -6c_1^4 c_2 + 6s_1 c_2^2 s_2 c_1^3 - c_2^3 + 6c_1^4 c_2^3 - 2s_1 s_2 c_1^3 + 3c_1^2 c_2 - c_1^2 c_2^3, \\
n_{12}^R &:= 6s_1 c_2 s_2 c_1 - 3c_1^2 + 7c_1^2 c_2^2 - 3c_2^2, \\
n_{22}^R &:= -c_1^3 c_2^2 - c_1^3 - 2s_1 c_2^3 s_2 + 6c_2^3 s_1 s_2 c_1^2 + 6c_1^3 c_2^4 - 6c_1 c_2^4 + 3c_1 c_2^2.
\end{aligned}
\tag{6.15}
$$

We note a similar conclusion for the denominator of the local pair correlation functions. In general, for the real power flow pair correlations, we find that the denominators of the local pair correlation functions are:

$$
\begin{aligned}
r_{11}^R := &\ -c_1^6 - 15c_1^2 c_2^4 + 42c_1^4 c_2^4 - 27c_1^4 c_2^6 + 15c_1^6 c_2^2 - 27c_1^6 c_2^4 \\
&+ 15c_1^2 c_2^6 + 6s_1 c_2 s_2 c_1^5 + 20s_1 c_2^3 s_2 c_1^3 - 15c_1^4 c_2^2 \\
&+ 6s_1 c_2^5 s_2 c_1 - c_2^6 + 13c_1^6 c_2^6 - 20s_1 c_2^3 s_2 c_1^5 \\
&- 20s_1 c_2^5 s_2 c_1^3 + 14c_1^5 c_2^5 s_1 s_2,
\end{aligned}
\tag{6.16}
$$

$$
\begin{aligned}
r_{12}^R := &\ -c_1^6 - 15c_1^2 c_2^4 + 42c_1^4 c_2^4 - 27c_1^4 c_2^6 + 15c_1^6 c_2^2 - 27c_1^6 c_2^4 \\
&+ 15c_1^2 c_2^6 + 6s_1 c_2 s_2 c_1^5 + 20s_1 c_2^3 s_2 c_1^3 - 15c_1^4 c_2^2 \\
&+ 6s_1 c_2^5 s_2 c_1 - c_2^6 + 13c_1^6 c_2^6 - 20s_1 c_2^3 s_2 c_1^5 \\
&- 20s_1 c_2^5 s_2 c_1^3 + 14c_1^5 c_2^5 s_1 s_2,
\end{aligned}
\tag{6.17}
$$

$$
\begin{aligned}
r_{22}^R := &\ -c_1^6 - 15c_1^2 c_2^4 + 42c_1^4 c_2^4 - 27c_1^4 c_2^6 + 15c_1^6 c_2^2 - 27c_1^6 c_2^4 \\
&+ 15c_1^2 c_2^6 + 6s_1 c_2 s_2 c_1^5 + 20s_1 c_2^3 s_2 c_1^3 - 15c_1^4 c_2^2 \\
&+ 6s_1 c_2^5 s_2 c_1 - c_2^6 + 13c_1^6 c_2^6 - 20s_1 c_2^3 s_2 c_1^5 \\
&- 20s_1 c_2^5 s_2 c_1^3 + 14c_1^5 c_2^5 s_1 s_2.
\end{aligned}
\tag{6.18}
$$

It is worth mentioning that the real network is well-behaved for the generic values of the parameters. Over the domain of the parameters $\{a_1, a_2\}$, we observe that the Gaussian fluctuations form a set of stable correlations if the determinant of the metric tensor, viz.,

$$g = \frac{8V^4 c_2^3 c_1^3}{R_0^2} \frac{n_g^R}{r_g^R}, \tag{6.19}$$

remains a positive function on the power factor surface $(M_2(R), g)$. We find that the numerator of the determinant of the metric tensor can be expressed as the trigonometric polynomial

$$n_g^R := 12 s_1 s_2 c_1^4 c_2^4 - 4 c_1^3 c_2^3 + 4 c_1^3 c_2 - c_1^5 c_2 - c_1 c_2^5 + 4 c_1 c_2^3 - c_1^4 s_1 s_2 - c_1^4 s_1 s_2 c_2^2$$
$$- c_2^4 s_1 s_2 c_1^2 - c_2^4 s_1 s_2 - 7 c_1^3 c_2^5 - 6 c_2^2 c_1^2 s_1 s_2 - 7 c_1^5 c_2^3 + 12 c_1^5 c_2^5. \tag{6.20}$$

As expected, the denominator of the determinant of the metric tensor also turns out to be given by a trigonometric polynomial, viz.,

$$\begin{aligned}
r_g^R := {}& -210 c_1^4 c_2^6 - 210 c_1^6 c_2^4 + 910 c_1^6 c_2^6 - c_1^{10} - c_2^{10} + 121 c_2^{10} c_1^{10} \\
& + 122 c_2^9 c_1^9 s_1 s_2 + 252 c_1^5 c_2^5 s_1 s_2 - 584 s_1 c_2^7 s_2 c_1^5 + 332 s_1 c_2^9 s_2 c_1^5 \\
& + 808 s_1 c_2^7 s_2 c_1^7 + 120 s_1 c_2^3 s_2 c_1^7 - 120 c_2^3 c_1^9 s_1 s_2 + 10 s_1 s_2 c_2 c_1^9 \\
& + 45 c_2^{10} c_1^2 - 250 c_2^{10} c_1^4 + 490 c_2^{10} c_1^6 - 405 c_2^{10} c_1^8 + 45 c_2^2 c_1^{10} \\
& - 120 s_1 c_2^9 s_2 c_1^3 + 10 s_1 c_2^9 s_2 c_1 - 584 s_1 c_2^5 s_2 c_1^7 - 344 s_1 c_2^9 s_2 c_1^7 \\
& - 344 s_1 c_2^7 s_2 c_1^9 + 332 c_2^5 c_1^9 s_1 s_2 + 120 s_1 c_2^7 s_2 c_1^3 - 250 c_2^4 c_1^{10} \\
& + 460 c_2^4 c_1^8 - 45 c_1^2 c_2^8 + 490 c_2^6 c_1^{10} - 1190 c_2^8 c_1^6 + 1180 c_2^8 c_1^8 \\
& - 405 c_2^8 c_1^{1} 0 - 1190 c_2^6 c_1^8 - 45 c_2^2 c_1^8 + 460 c_2^8 c_1^4. \tag{6.21}
\end{aligned}$$

Here, the behavior of the determinant of the metric tensor shows that such a real power flow becomes unstable for specific values of the parameters. For generic R_0 and V, the nature of the determinant of the metric tensor is described by (6.19). It is worth mentioning further that the electric networks become unstable in the limit of vanishing $\{a_1, a_2\}$.

In order to explain the nature of the transformation of the $\{a_1, a_2\}$ forming the intrinsic surface, let us explore the functional behavior of the associated scalar curvature. Our computation shows that the scalar curvature reduces to the form

$$R = \frac{1}{4 R_0 c_1^2 c_2^2 V^2} \frac{n_R^R}{r_R^R}. \tag{6.22}$$

The numerator of the scalar curvature is given by the trigonometric polynomial expression

$$\begin{aligned}
n_R^R := {}& -42 c_1^4 c_2^6 - 42 c_1^6 c_2^4 + 444 c_1^6 c_2^6 + 3 c_1^{10} + 3 c_2^{10} + 702 c_2^{10} c_1^{10} \\
& + 702 c_2^9 c_1^9 s_1 s_2 + 84 c_1^5 c_2^5 s_1 s_2 - 280 s_1 c_2^7 s_2 c_1^5 + 356 s_1 c_2^9 s_2 c_1^5 \\
& + 1080 s_1 c_2^7 s_2 c_1^7 - 24 s_1 c_2^3 s_2 c_1^7 + 24 c_2^3 c_1^9 s_1 s_2 - 18 s_1 s_2 c_2 c_1^9
\end{aligned}$$

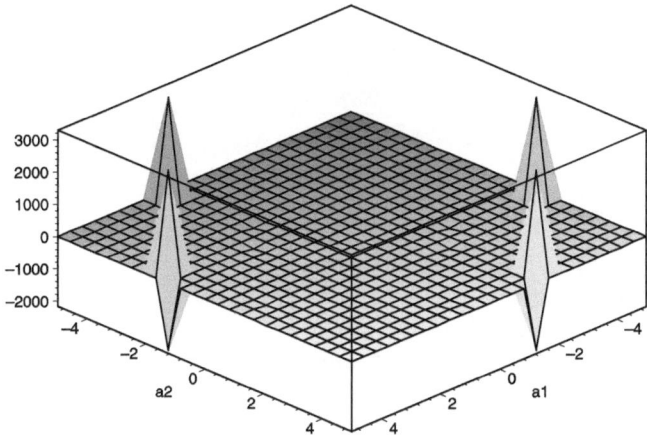

Fig. 6.1 Determinant of the metric tensor plotted as a function of the power factors a_1 and a_2, describing real power fluctuations in electrical networks

$$
\begin{aligned}
&- 38c_2^{10}c_1^2 - 70c_2^{10}c_1^4 + 752c_2^{10}c_1^6 - 1349c_2^{10}c_1^8 - 38c_2^2c_1^10 \\
&+ 24s_1c_2^9s_2c_1^3 - 18s_1c_2^9s_2c_1 - 280s_1c_2^5s_2c_1^7 - 1000s_1c_2^9s_2c_1^7 \\
&- 1000s_1c_2^7s_2c_1^9 + 356c_2^5c_1^9s_1s_2 - 24s_1c_2^7s_2c_1^3 - 70c_2^4c_1^{10} \\
&+ 72c_2^4c_1^8 + 39c_2^2c_1^8 + 752c_2^6c_1^10 - 1082c_2^8c_1^6 + 2288c_2^8c_1^8 \\
&- 1349c_2^8c_1^{10} - 1082c_2^6c_1^8 + 39c_2^8c_1^8 + 72c_2^8c_1^4, \quad (6.23)
\end{aligned}
$$

while the denominator of the scalar curvature can be expressed as the trigonometric polynomial

$$
\begin{aligned}
r_R^R := &-c_1^6 + 6s_1c_2s_2c_1^5 + 20s_1c_2^3s_2c_1^3 + 6s_1c_2^5s_2c_1 + 8s_1c_2^3s_2c_1^5 \\
&+ 8s_1c_2^5s_2c_1^3 - 15c_1^2c_2^4 + 6c_1^4c_2^4 + 53c_1^4c_2^6 + 4c_1^6c_2^2 + 53c_1^6c_2^4 \\
&+ 4c_1^2c_2^6 - 15c_1^4c_2^2 - 48c_1^6c_2^6 + c_1^8 + c_2^8 - 70c_1^5c_2^5s_1s_2 - c_2^6 \\
&- 12s_1c_2^7s_2c_1^5 + 72s_1c_2^5s_2c_1^7 - 12s_1c_2^3s_2c_1^7 - 12s_1c_2^5s_2c_1^7 \\
&- 12s_1c_2^7s_2c_1^3 - 11c_2^4c_1^8 + 2c_1^2c_2^8 - 48c_2^8c_1^6 + 72c_2^8c_1^8 \\
&- 48c_2^6c_1^8 + 2c_2^2c_1^8 - 11c_2^8c_1^4. \quad (6.24)
\end{aligned}
$$

We find that a typical real power network is globally correlated over all the generic Gaussian fluctuations of the parameters $\{a_1, a_2\}$, unless $n_R^R = 0$. For the above real power networks, we observe that the scalar curvature diverges in the limit $r_R^R = 0$, signature of a global instability on the $\{a_1, a_2\}$ surface. Thus, the intrinsic geometric analysis shows that a real power network is interacting and locally stable over the surface of fluctuation, if the network parameters $\{a_1, a_2\}$ are properly chosen.

For the choices $V = 1$ and $R_0 = 1$, Fig. 6.1 shows the determinant of the metric tensor. These plots explicate the nature of the stability of real power flows in power networks. The corresponding plot for the scalar curvature is given in Fig. 6.2. This

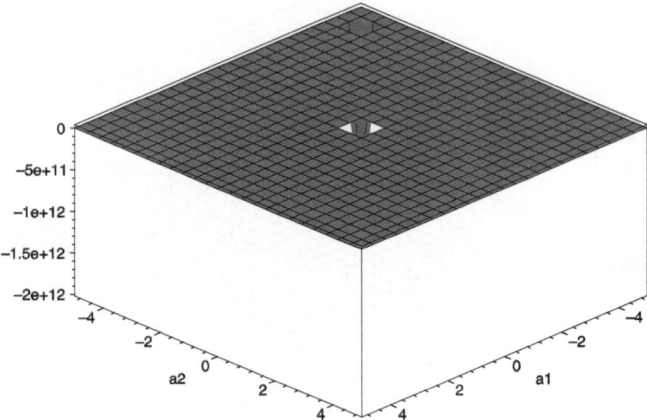

Fig. 6.2 Curvature scalar plotted as a function of the power factors a_1 and a_2, describing the real power fluctuations in electrical networks

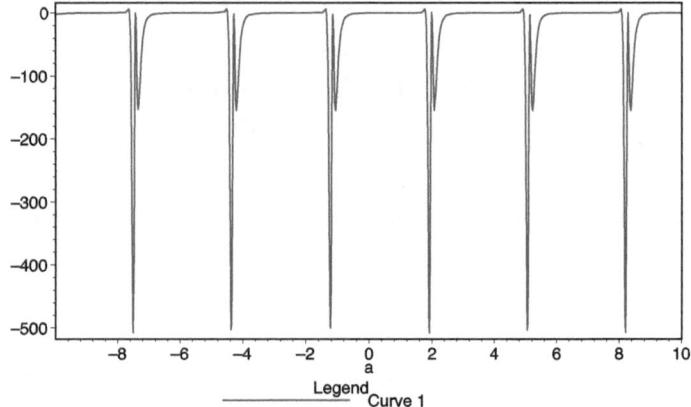

Fig. 6.3 Determinant of the metric tensor plotted as a function of the equal power factors $a :=$ $a_1 = a_2$, describing real power fluctuations in electrical networks

plot shows the global nature of the real power flow in an electrical network under the effects of Gaussian fluctuations of the parameters.

For equal phases, viz., $a_1 = a$ and $a_2 = a$, the surface plots of the determinant of the metric tensor and scalar curvature are shown in Figs. 6.3 and 6.4, respectively. We observe that the stability of real power networks exists in certain bands. This follows from the fact that the instability is present only for a specific set of equal values of the parameters. Interestingly, for a network with limiting equal phases, the limiting scalar curvature simplifies to the shape illustrated. In particular, the corresponding peaks in Figs. 6.2 and 6.4 showing the scalar curvature indicate graphically the nature

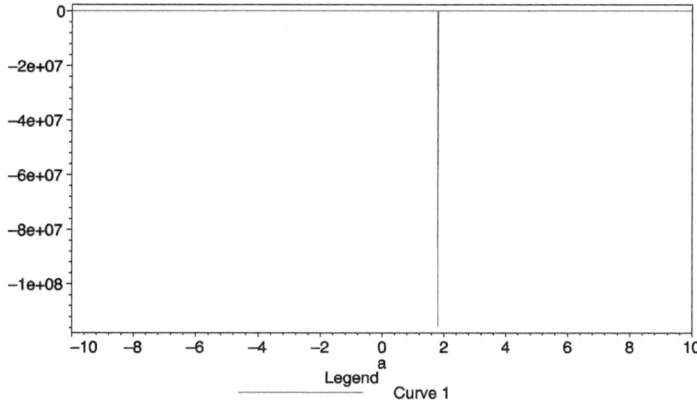

Fig. 6.4 Curvature scalar plotted as a function of the equal power factors $a := a_1 = a_2$, describing the real fluctuations in electrical networks

of the global instability in real power networks. Physically, the peaks in the curvature show the presence of non-trivial interactions in the network.

6.4 Imaginary Power Flow

In the present section, we analyze the nature of an ensemble of fluctuating electrical networks generated by a pair a_1, a_2. To focus on the most general case, we choose the variables a_1, a_2 as a function of L, C, and r for the given network. By (6.2), when the imaginary power

$$Q(a_1, a_2) := \frac{V^2}{R_0} \frac{\tan a_1 - \tan a_2}{1 + (\tan a_1 - \tan a_2)^2} \tag{6.25}$$

is allowed to fluctuate as a function of the a_1, a_2, we may again exploit the definition of the Hessian function $\text{Hess}(Q(a_1, a_2))$ of the imaginary power. Here, the components of the metric tensor are given by

$$g_{a_1 a_1} = -\frac{2c_2^2 V^2}{R_0} \frac{n_{11}^I}{r_{11}^I}, \quad g_{a_1 a_2} = \frac{2c_2^2 c_1^2 V^2}{R_0} \frac{n_{12}^I}{r_{12}^I}, \quad g_{a_2 a_2} = -\frac{2c_1^2 V^2}{R_0} \frac{n_{22}^I}{r_{22}^I}. \tag{6.26}$$

In this case, the numerators of the local pair correlation functions are expressed by the following trigonometric polynomials:

$$n_{11}^I := -5c_2^3 c_1^2 s_2 - s_1 c_1^3 + 8c_2^2 s_1 c_1^3 - 4c_2 s_2 c_1^4 + 3c_2 s_2 c_1^2$$
$$- 3c_2^2 s_1 c_1 + 8c_1^4 s_2 c_2^3 - 7s_1 c_1^3 c_2^4 + s_2 c_2^3 + c_2^4 s_1 c_1 \,,$$
$$n_{12}^I := c_2^3 s_1 + 3c_1^2 s_1 c_2 - c_1^3 s_2 - 7c_2^3 c_1^2 s_1 + 7c_2^2 s_2 c_1^3 - 3c_2^3 s_2 c_1 \,, \tag{6.27}$$

$$n_{22}^I := -8c_2^3c_1^2s_2 - s_1c_1^3 + 5c_2^2s_1c_1^3 - c_2s_2c_1^4 + 3c_2s_2c_1^2$$
$$- 3c_2^2s_1c_1 + 7c_1^4s_2c_2^3 - 8s_1c_1^3c_2^4 + s_2c_2^3 + 4c_2^4s_1c_1,$$

while the denominators of the local pair correlation functions have the following trigonometric expressions:

$$r_{11}^I := 13c_1^6c_2^6 + 14c_1^5c_2^5s_1s_2 + 6s_1c_2^5s_2c_1 + 15c_1^2c_2^6$$
$$+ 6s_1c_2s_2c_1^5 - 20c_1^3c_2^5s_1s_2 - 20c_1^5c_2^3s_1s_2 - 15c_1^2c_2^4$$
$$+ 20c_1^3c_2^3s_1s_2 - 27c_1^4c_2^6 + 15c_1^6c_2^2 - 27c_1^6c_2^4 + 42c_1^4c_2^4$$
$$- 15c_1^4c_2^2 - c_2^6 - c_1^6, \tag{6.28}$$

$$r_{12}^I := 13c_1^6c_2^6 + 14c_1^5c_2^5s_1s_2 + 6s_1c_2^5s_2c_1 + 15c_1^2c_2^6$$
$$+ 6s_1c_2s_2c_1^5 - 20c_1^3c_2^5s_1s_2 - 20c_1^5c_2^3s_1s_2 - 15c_1^2c_2^4$$
$$+ 20c_1^3c_2^3s_1s_2 - 27c_1^4c_2^6 + 15c_1^6c_2^2 - 27c_1^6c_2^4 + 42c_1^4c_2^4$$
$$- 15c_1^4c_2^2 - c_2^6 - c_1^6, \tag{6.29}$$

$$r_{22}^I := 13c_1^6c_2^6 + 14c_1^5c_2^5s_1s_2 + 6s_1c_2^5s_2c_1 + 15c_1^2c_2^6$$
$$+ 6s_1c_2s_2c_1^5 - 20c_1^3c_2^5s_1s_2 - 20c_1^5c_2^3s_1s_2 - 15c_1^2c_2^4$$
$$+ 20c_1^3c_2^3s_1s_2 - 27c_1^4c_2^6 + 15c_1^6c_2^2 - 27c_1^6c_2^4 + 42c_1^4c_2^4$$
$$- 15c_1^4c_2^2 - c_2^6 - c_1^6. \tag{6.30}$$

It follows that the pure pair correlations $\{g_{11}, g_{22}\}$ between the parameters $\{a_1, a_2\}$ remain positive, which is the same as for the flow of the real power. A straightforward computation further demonstrates the global nature of the parametric fluctuations. In fact, we find that the determinant of the metric tensor reduces to the expression

$$g = -\frac{V^4 c_2^3 c_1^3}{R_0^2} \frac{n_g^I}{r_g^I}. \tag{6.31}$$

In this case, the numerator of the determinant of the metric tensor is given by the following trigonometric polynomial:

$$n_g^I := 10c_1^6c_2^2 + 37c_1^5c_2^5s_1s_2 - 3c_1^4c_2^2 - 3c_1^2c_2^4 - 39c_1^6c_2^4 + 28c_1^4c_2^4$$
$$+ 10c_1^2c_2^6 - 39c_1^4c_2^6 + 3s_1c_2^5s_2c_1 - 20c_1^3c_2^5s_1s_2 - 20c_1^5c_2^3s_1s_2$$
$$+ 3s_1c_2s_2c_1^5 + 2c_1^3c_2^3s_1s_2 + 38c_1^6c_2^6 - c_2^6 - c_1^6. \tag{6.32}$$

Explicitly, we find that the denominator of the determinant of the metric tensor can be written

$$r_g^I := 252c_1^5c_2^5s_1s_2 + 910c_1^6c_2^6 - 210c_1^4c_2^6 - 210c_1^6c_2^4$$
$$- c_1^{10} - c_2^{10} + 122c_1^9c_2^9s_1s_2 + 121c_1^{10}c_2^{10} + 10s_1c_2^9s_2c_1$$
$$- 120s_1c_2^9s_2c_1^3 - 120c_1^9c_2^3s_1s_2 + 332c_1^3c_2^9s_1s_2 + 332c_1^9c_2^5s_1s_2$$

$$-344c_1^9c_2^7s_1s_2 - 344c_1^7c_2^9s_1s_2 + 490c_1^6c_2^{10} - 405c_1^8c_2^{10} - 250c_1^{10}c_2^4$$
$$+490c_1^{10}c_2^6 + 1180c_2^8c_1^8 - 1190c_2^6c_1^8 - 1190c_2^8c_1^6 - 250c_1^4c_2^{10} - 405c_1^{10}c_2^8$$
$$-45c_1^8c_2^2 - 45c_2^8c_1^2 + 460c_1^8c_2^4 + 460c_2^8c_1^4 + 45c_1^{10}c_2^2 + 45c_2^{10}c_1^2$$
$$+120c_1^7c_2^3s_1s_2 - 584c_1^7c_2^5s_1s_2 + 120c_1^3c_2^7s_1s_2 - 584c_1^5c_2^7s_1s_2$$
$$+10s_1s_2c_1^9c_2 + 808c_1^7c_2^7s_1s_2. \tag{6.33}$$

It is not difficult to compute the exact expression for the scalar curvature describing the global parametric intrinsic correlations. In particular, we find that the scalar curvature reduces to the form

$$R = -\frac{R_0}{2c_1^2c_2^2V^2}\,\frac{n_R^{(1)\mathrm{I}} + n_R^{(2)\mathrm{I}}}{r_R^{\mathrm{I}}}. \tag{6.34}$$

In the above equation, the numerator of the scalar curvature has the trigonometric polynomial expression

$$\begin{aligned}
n_R^{(1)\mathrm{I}} := {}& -297s_1c_1^5c_2^8 + 2235s_1c_1^5c_2^{10} + 4770s_1c_1^7c_2^8 - 15596s_1c_1^7c_2^{10} \\
& -19538s_1c_1^9c_2^8 + 45624s_1c_1^9c_2^{10} - 7857s_1c_1^{11}c_2^6 - 3954s_1c_1^5c_2^{12} \\
& +19890s_1c_1^7c_2^{12} - 49482s_1c_1^9c_2^{12} - 947s_2c_1^4c_2^{11} + 59571s_1c_1^{11}c_2^{12} \\
& +121s_2c_1^{10}c_2^3 - 2235s_2c_1^{10}c_2^5 + 30498s_1c_1^{11}c_2^8 - 3653s_2c_1^6c_2^9 \\
& -11s_2c_1^{14}c_2 + 15596s_2c_1^{10}c_2^7 - 45624s_2c_1^{10}c_2^9 + 60771s_2c_1^{10}c_2^{11} \\
& -355s_2c_1^{12}c_2^3 + 3954s_2c_1^{12}c_2^5 - 19890s_2c_1^{12}c_2^7 + 49482s_2c_1^{12}c_2^9 \\
& -59571s_2c_1^{12}c_2^{11} - 60771s_1c_1^{11}c_2^{10} + 19538s_2c_1^8c_2^9 + 7857s_2c_1^6c_2^{11} \\
& -30498s_2c_1^8c_2^{11} + 2080s_1c_1^5c_2^{14} - 8902s_1c_1^7c_2^{14} + 20139s_1c_1^9c_2^{14} \\
& -22407s_1c_1^{11}c_2^{14} + 8902s_2c_1^{14}c_2^7 - 20139s_2c_1^{14}c_2^9 + 22407s_2c_1^{14}c_2^{11}, \tag{6.35}
\end{aligned}$$

$$\begin{aligned}
n_R^{(2)\mathrm{I}} := {}& 11s_1c_1c_2^{14} - 2080s_2c_1^{14}c_2^5 - 243s_1c_1^3c_2^{14} - 9450s_2c_2^{13}c_1^{14} \\
& +61s_1c_1^{13}c_2^2 - 812s_1c_1^{13}c_2^4 + 5044s_1c_1^{13}c_2^6 - 16769s_1c_1^{13}c_2^8 \\
& +30165s_1c_1^{13}c_2^{10} - 27138s_1c_1^{13}c_2^{12} + 9450s_1c_1^{13}c_2^{14} \\
& -5044s_2c_1^6c_2^{13} - 61s_2c_1^2c_2^{13} + 220s_2c_1^4c_2^9 - 121s_1c_1^3c_2^{10} \\
& +947s_1c_1^{11}c_2^4 + 355s_1c_1^3c_2^{12} - 45s_1c_1^{11}c_2^2 + 27138c_1^{12}c_2^{13}s_2 \\
& +330s_2c_1^6c_2^7 + 3653s_1c_1^9c_2^6 - 220s_1c_1^9c_2^4 - 10s_1c_1c_2^{12} \\
& +297s_2c_1^8c_2^5 - 330s_1c_1^7c_2^6 + 45s_2c_1^2c_2^{11} + 10s_2c_1^{12}c_2 \\
& -4770s_2c_1^8c_2^7 - 30165s_2c_1^{10}c_2^{13} + 812s_2c_1^4c_2^{13} - s_1c_1^{13} \\
& +16769s_2c_1^8c_2^{13} + s_2c_2^{13} + 243s_2c_1^{14}c_2^3. \tag{6.36}
\end{aligned}$$

The function r_R^{I} appearing in the denominator of the scalar curvature can be written as the polynomial

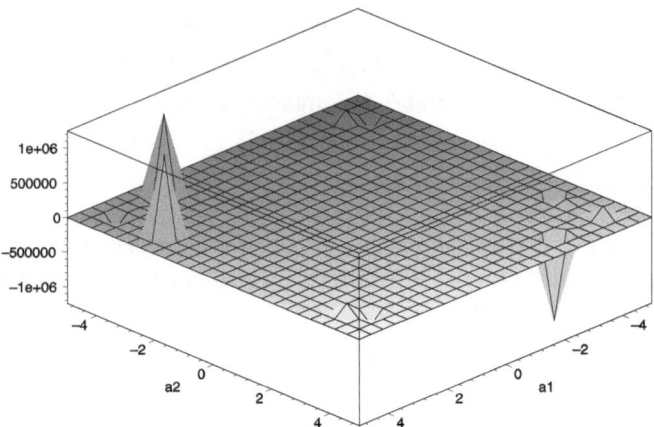

Fig. 6.5 Determinant of the metric tensor plotted as a function of the power factors a_1 and a_2, describing the imaginary power fluctuations in electrical networks

$$
\begin{aligned}
r_R^I := {} & 42c_1^6 c_2^6 + c_2^{12} + c_1^{12} + 5344c_1^9 c_2^9 s_1 s_2 + 10447c_1^{10} c_2^{10} + 4383c_1^{12} c_2^8 \\
& + 100c_1^3 c_2^{11} s_1 s_2 - 708c_1^5 c_2^{11} s_1 s_2 + 100c_1^{11} c_2^3 s_1 s_2 + 2528c_1^7 c_2^{11} s_1 s_2 \\
& - 4406c_1^9 c_2^{11} s_1 s_2 + 2528c_1^{11} c_2^7 s_1 s_2 - 4406c_1^{11} c_2^9 s_1 s_2 + 2812c_1^{11} c_2^{11} s_1 s_2 \\
& - 6c_1 c_2^{11} s_1 s_2 - 708c_1^{11} c_2^5 s_1 s_2 - 6c_1^{11} c_2 s_1 s_2 - 29c_1^2 c_2^{12} - 29c_1^{12} c_2^2 \\
& + 4383c_1^8 c_2^{12} - 5813c_1^{10} c_2^{12} + 2813c_1^{12} c_2^{12} - 1598c_1^6 c_2^{12} - 1598c_1^{12} c_2^6 \\
& - 5813c_1^{12} c_2^{10} + 307c_1^4 c_2^{12} + 307c_1^{12} c_2^4 - 22s_1 c_2^9 s_2 c_1^3 - 22c_1^9 c_2^3 s_1 s_2 \\
& + 368c_1^5 c_2^9 s_1 s_2 + 368c_1^9 c_2^5 s_1 s_2 - 2132c_1^9 c_2^7 s_1 s_2 - 2132c_1^7 c_2^9 s_1 s_2 \\
& + 1826c_2^6 c_1^{10} - 6442c_1^8 c_2^{10} - 257c_1^{10} c_2^4 + 1826c_1^{10} c_2^6 + 2822c_2^8 c_1^8 \\
& - 482c_2^6 c_1^8 - 482c_2^8 c_1^6 - 257c_1^4 c_2^{10} - 6442c_1^{10} c_2^8 + 27c_1^8 c_2^4 + 27c_2^8 c_1^4 \\
& + 15c_1^{10} c_2^2 + 15c_2^{10} c_1^2 - 36c_1^7 c_2^5 s_1 s_2 - 36c_1^5 c_2^7 s_1 s_2 + 472c_1^7 c_2^7 s_1 s_2. \quad (6.37)
\end{aligned}
$$

Consequently, we may easily analyze the underlying stability properties for specific consideration of variable power factors. As in the case of real power flow, the global nature of the scalar curvature and associated phase transitions can be determined over the range of power factors describing the relevant network.

As in the last section, we shall focus our attention on the same electrical network and on the specific values $V = 1$ and $R_0 = 1$. The determinant of the metric tensor shown in Fig. 6.5 describes the phenomenological properties of Gaussian imaginary power fluctuations. The scalar curvature in Fig. 6.6 exhibits a couple of antisymmetric fluctuations for the imaginary network power flow.

For equal values $a_1 = a$ and $a_2 = a$, we note in Fig. 6.7 that the system acquires a couple of locally chaotic fluctuations. Here, we find the surprising fact that the imaginary power flow has a different geometric nature, and the scalar curvature turns out to be zero in the limit of equal phases for the imaginary power flow. It is worth mentioning that such a power flow is globally non-interacting.

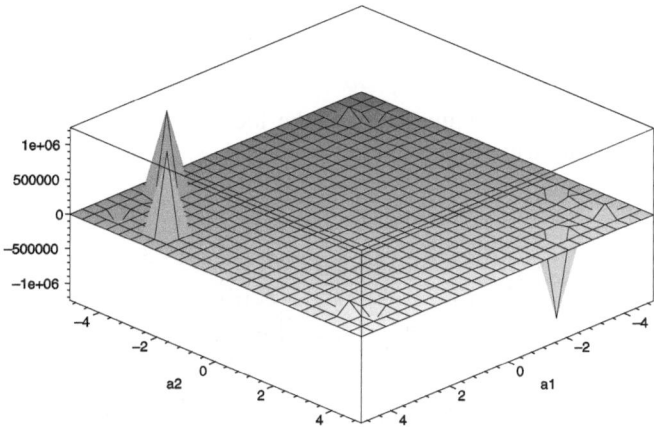

Fig. 6.6 Curvature scalar plotted as a function of the power factors a_1 and a_2, describing the imaginary power fluctuations in electrical networks

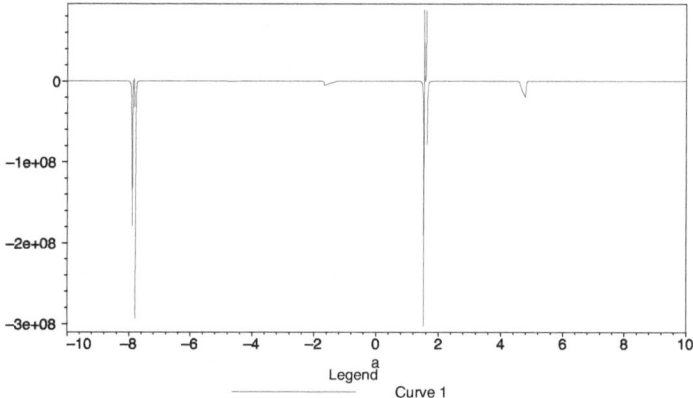

Fig. 6.7 Curvature scalar plotted as a function of the power factors a_1 and a_2, describing the imaginary power fluctuations in electrical networks

6.5 Complex Power Flow

In order to further understand the nature of generic electrical networks, we now consider linear combinations of the real and imaginary power flows in the network. The associated joint power flow F takes the form

$$F(a_1, a_2) := \frac{V^2}{R_0} \frac{1 + \tan a_1 - \tan a_2}{1 + (\tan a_1 - \tan a_2)^2} . \tag{6.38}$$

To obtain the components of the metric tensor in the power space, we apply the definition of the Hessian matrix as indicated previously, whence the metric tensor components are found to have the form:

$$g_{a_1 a_1} = \frac{2c_2^2 V^2}{R_0} \frac{n_{11}^C}{r_{11}^C}, \quad g_{a_1 a_2} = -\frac{2c_2 c_1 V^2}{R_0} \frac{n_{12}^C}{r_{12}^C}, \quad g_{a_2 a_2} = \frac{2c_1^2 V^2}{R_0} \frac{n_{22}^C}{r_{22}^C}. \quad (6.39)$$

In the above equation, the numerators of the local pair correlation functions are expressed by the following trigonometric polynomials:

$$\begin{aligned}
n_{11}^C := & -c_2^3 s_2 + s_1 c_1^3 + 3c_1^2 c_2^2 - 2c_2 s_1 s_2 c_1^3 + 5c_1^2 c_2^3 s_2 \\
& + 4c_1^4 c_2 s_2 - c_1 c_2^4 s_1 + 6c_1^3 c_2^3 s_1 s_2 - 3c_1^2 c_2 s_2 \\
& + 7c_1^3 c_2^4 s_1 - 8c_1^4 c_2^3 s_2 + 3c_1 c_2^2 s_1 + 6c_1^4 c_2^4 \\
& - 6c_1^4 c_2^2 - 8c_1^3 c_2^2 s_1 - c_2^4 c_1^2 - c_2^4,
\end{aligned} \quad (6.40)$$

$$\begin{aligned}
n_{12}^C := & \; 6c_1^2 c_2^3 s_1 s_2 - 3c_1^3 c_2 - 3c_1^2 s_1 c_2 + c_1^3 s_2 - 3c_1 c_2^3 \\
& + 7c_1^3 c_2^3 + 7s_1 c_2^3 c_1^2 - 7s_2 c_1^3 c_2^2 - c_2^3 s_1 + 3c_2^3 s_2 c_1,
\end{aligned} \quad (6.41)$$

$$\begin{aligned}
n_{22}^C := & -c_2^3 s_2 + s_1 c_1^3 + 3c_1^2 c_2^2 + 8c_1^2 c_2^3 s_2 + c_1^4 c_2 s_2 \\
& - 4c_1 c_2^4 s_1 + 6c_1^3 c_2^3 s_1 s_2 - 3c_1^2 c_2 s_2 + 8c_1^3 c_2^4 s_1 \\
& - 7c_1^4 c_2^3 s_2 + 3c_1 c_2^2 s_1 + 6c_1^4 c_2^4 - c_1^4 - c_1^4 c_2^2 \\
& - 5c_1^3 c_2^2 s_1 - 6c_2^4 c_1^2 - 2c_2^3 s_1 s_2 c_1.
\end{aligned} \quad (6.42)$$

The denominators of the local pair correlation functions reduce to the following trigonometric polynomials:

$$\begin{aligned}
r_{11}^C := & -20c_2^3 c_1^5 s_1 s_2 - c_2^6 - c_1^6 - 27c_2^4 c_1^6 + 14c_1^5 c_2^5 s_1 s_2 \\
& + 13c_1^6 c_2^6 + 6s_1 c_2^5 s_2 c_1 + 15c_1^2 c_2^6 - 27c_1^4 c_2^6 \\
& - 20c_1^3 c_2^5 s_1 s_2 + 6c_2 c_1^5 s_1 s_2 + 20c_1^3 c_2^3 s_1 s_2 \\
& + 42c_1^4 c_2^4 + 15c_1^6 c_2^2 - 15c_1^4 c_2^2 - 15c_2^4 c_1^2,
\end{aligned} \quad (6.43)$$

$$\begin{aligned}
r_{12}^C := & -20c_2^3 c_1^5 s_1 s_2 - c_2^6 - c_1^6 - 27c_2^4 c_1^6 + 14c_1^5 c_2^5 s_1 s_2 \\
& + 13c_1^6 c_2^6 + 6s_1 c_2^5 s_2 c_1 + 15c_1^2 c_2^6 - 27c_1^4 c_2^6 \\
& - 20c_1^3 c_2^5 s_1 s_2 + 6c_2 c_1^5 s_1 s_2 + 20c_1^3 c_2^3 s_1 s_2 \\
& + 42c_1^4 c_2^4 + 15c_1^6 c_2^2 - 15c_1^4 c_2^2 - 15c_2^4 c_1^2,
\end{aligned} \quad (6.44)$$

$$\begin{aligned}
r_{22}^C := & -20c_2^3 c_1^5 s_1 s_2 - c_2^6 - c_1^6 - 27c_2^4 c_1^6 + 14c_1^5 c_2^5 s_1 s_2 \\
& + 13c_1^6 c_2^6 + 6s_1 c_2^5 s_2 c_1 + 15c_1^2 c_2^6 - 27c_1^4 c_2^6 \\
& - 20c_1^3 c_2^5 s_1 s_2 + 6c_2 c_1^5 s_1 s_2 + 20c_1^3 c_2^3 s_1 s_2 \\
& + 42c_1^4 c_2^4 + 15c_1^6 c_2^2 - 15c_1^4 c_2^2 - 15c_2^4 c_1^2.
\end{aligned} \quad (6.45)$$

The determinant of the metric tensor turns out to be a rational polynomial function in the parameters $\{a_1, a_2\}$. In compact notation, it is given by

$$g = -\frac{4V^4 c_2^2 c_1^2}{R_0^2} \frac{n_g^C}{r_g^C}. \tag{6.46}$$

Here, the numerator of the determinant of the metric tensor is given by the trigonometric polynomial

$$
\begin{aligned}
n_g^C := & - 25c_2^4 c_1^6 - 18c_2^3 c_1^5 s_1 s_2 - c_2^6 - c_1^6 + 13c_1^5 c_2^5 s_1 s_2 \\
& - c_1^5 s_1 + s_2 c_2^5 + 5s_1 c_2^5 s_2 c_1 - 18c_1^3 c_2^5 s_1 s_2 + 5c_2 c_1^5 s_1 s_2 \\
& + 12c_1^2 c_2^6 - 25c_1^4 c_2^6 + 12c_2^6 c_2^2 + 5c_1^4 c_2 s_2 - 5c_1 c_2^4 s_1 \\
& - 10c_1^3 c_2^2 s_1 - 11c_1^4 c_2^2 - 11c_2^4 c_1^2 + 10c_1^2 c_2^3 s_2 - 60s_1 c_2^6 c_1^5 \\
& + 60s_2 c_2^5 c_1^6 - 20c_1^6 c_2^3 s_2 + 20c_1^3 c_2^6 s_1 - 2c_1^5 c_2^2 s_1 - 47s_2 c_2^5 c_1^4 \\
& + 2s_2 c_2^5 c_1^2 + 47c_1^5 c_2^4 s_1 + 14c_1^3 c_2^3 s_1 s_2 + 14c_1^6 c_2^6 + 36c_1^4 c_2^4 \\
& - 10c_1^4 c_2^3 s_2 + 10c_1^3 c_2^4 s_1, \tag{6.47}
\end{aligned}
$$

while the denominator of the metric determinant is given by the trigonometric polynomial

$$
\begin{aligned}
r_g^C := & - 210c_2^4 c_1^6 + 252c_1^5 c_2^5 s_1 s_2 + 460c_1^4 c_2^8 - 1190c_1^6 c_2^8 + 460c_1^8 c_2^4 \\
& - 1190c_1^8 c_2^6 - 45c_1^2 c_2^8 - 45c_1^8 c_2^2 + 45c_1^2 c_2^{10} - 250c_1^4 c_2^{10} + 490c_1^6 c_2^{10} \\
& - 405c_1^8 c_2^{10} - 405c_1^{10} c_2^8 + 490c_1^{10} c_2^6 + 45c_1^{10} c_2^2 - 250c_1^{10} c_2^4 \\
& - 584c_1^5 c_2^7 s_1 s_2 + 120c_1^7 c_2^3 s_1 s_2 + 120c_1^3 c_2^7 s_1 s_2 - 584c_1^7 c_2^5 s_1 s_2 \\
& + 10s_1 s_2 c_1^9 c_2 + 10s_1 c_2^9 s_2 c_1 - 344c_1^7 c_2^5 s_1 s_2 - 120c_1^9 c_2^3 s_1 s_2 \\
& + 332c_1^9 c_2^5 s_1 s_2 - 344c_2^9 c_1^7 s_1 s_2 - 120s_1 c_2^9 s_2 c_1^3 + 332s_1 c_2^9 s_2 c_1^5 \\
& + 1180c_1^8 c_2^8 - 210c_1^4 c_2^6 - c_1^{10} + 808c_1^7 c_2^7 s_1 s_2 + 910c_1^6 c_2^6 \\
& - c_2^{10} + 121c_1^{10} c_2^{10} + 122c_1^9 c_2^9 s_1 s_2. \tag{6.48}
\end{aligned}
$$

We see that the determinant of the metric tensor remains nonzero in the space of power factors and thus defines a non-degenerate intrinsic geometry on the phase fluctuation surface.

Finally, we easily obtain the underlying scalar curvature, which also has a rational polynomial form. It turns out that the scalar curvature can be reduced to the form

$$R = \frac{R_0}{4c_1^2 c_2^2 V^2} \frac{n_R^{(1)C} + n_R^{(2)C} + n_R^{(3)C} + n_R^{(4)C}}{r_R^{(1)C} + r_R^{(2)C} + r_R^{(3)C}}. \tag{6.49}$$

Computation shows that the numerator of the scalar curvature is given by the following trigonometric polynomial expressions:

$$
\begin{aligned}
n_R^{(1)C} :=& -495c_1^4c_2^8 - 234c_1^6c_2^8 - 495c_1^8c_2^4 - 234c_1^8c_2^6 - 66c_1^2c_2^{10} + 162c_1^4c_2^{10} \\
& + 13566c_1^6c_2^{10} - 91736c_1^8c_2^{10} - 91736c_1^{10}c_2^8 + 13566c_1^{10}c_2^6 - 66c_1^{10}c_2^2 \\
& + 162c_1^{10}c_2^4 + 792c_1^5c_2^7s_1s_2 + 792c_1^7c_2^5s_1s_2 - 21064c_1^7c_2^9s_1s_2 + 220c_1^9c_2^3s_1s_2 \\
& + 492c_1^9c_2^5s_1s_2 - 21064c_1^9c_2^7s_1s_2 + 220s_1c_2^9s_2c_1^3 + 492s_1c_2^9s_2c_1^5 \\
& + 25506c_1^8c_2^8 - c_1^{12} - c_2^{12} + 2c_2^{14} + 1152c_1^7c_2^7s_1s_2 + 2c_1^{14} - 924c_1^6c_2^6 \\
& + 175360c_1^{11}c_2^{11}s_1s_2 + 233534c_1^{10}c_2^{10} + 85296c_1^9c_2^9s_1s_2 + 186s_1c_1^{13}c_2^2 \\
& - 1960s_1c_1^{13}c_2^4 + 5512s_1c_1^{13}c_2^6 + 3214s_1c_1^{13}c_2^8 - 34758s_1c_1^{13}c_2^{10} \\
& + 49084s_1c_1^{13}c_2^{12} - 2054s_2c_1^4c_2^{11} + 10386s_2c_1^6c_2^{11} - 9732s_2c_1^8c_2^{11} \\
& - 33034s_2c_1^{10}c_2^{11} + 72538s_2c_1^{12}c_2^{11} - 38450s_2c_1^{14}c_2^{11} + 14436s_2c_1^8c_2^9, \quad (6.50)
\end{aligned}
$$

$$
\begin{aligned}
n_R^{(2)C} :=& \, 90s_2c_1^2c_2^{11} - 5796s_2c_1^8c_2^7 + 594s_2c_1^8c_2^5 - 3878s_2c_1^{10}c_2^5 \\
& + 14424s_2c_1^{10}c_2^7 - 6944s_2c_1^{10}c_2^9 + 6996s_2c_1^{12}c_2^5 - 12292s_2c_1^{12}c_2^7 \\
& - 20348s_2c_1^{12}c_2^9 - 3600s_2c_1^{14}c_2^5 + 2428s_2c_1^{14}c_2^7 + 18194s_2c_1^{14}c_2^9 \\
& - 14436s_1c_1^9c_2^8 + 6944s_1c_1^9c_2^{10} + 20348s_1c_1^9c_2^{12} - 18194s_1c_1^9c_2^{14} \\
& - 10386s_1c_1^{11}c_2^6 + 9732s_1c_1^{11}c_2^8 + 33034s_1c_1^{11}c_2^{10} - 72538s_1c_1^{11}c_2^{12} \\
& + 38450s_1c_1^{11}c_2^{14} - 5074s_2c_1^6c_2^9 - 902s_2c_1^{12}c_2^3 + 694s_2c_1^{14}c_2^3 \\
& + 5074s_1c_1^9c_2^6 + 440s_2c_1^4c_2^9 - 30s_2c_1^{14}c_2 + 2054s_1c_1^{11}c_2^4 \\
& + 5796s_1c_1^7c_2^8 - 14424s_1c_1^7c_2^{10} - 2428s_1c_1^7c_2^{14} - 440s_1c_1^9c_2^4 \\
& - 660s_1c_1^7c_2^6 + 12292s_1c_1^7c_2^{12} + 902s_1c_1^3c_2^{12} - 6996s_1c_1^5c_2^{12}, \quad (6.51)
\end{aligned}
$$

$$
\begin{aligned}
n_R^{(3)C} :=& + 3878s_1c_1^5c_2^{10} + 3600s_1c_1^5c_2^{14} - 694s_1c_1^3c_2^{14} + 242s_2c_1^{10}c_2^3 \\
& - 594s_1c_1^5c_2^8 + 660s_2c_1^6c_2^7 - 242s_1c_1^3c_2^{10} - 20s_1c_1c_2^{12} \\
& + 30s_1c_1c_2^{14} + 20s_2c_1^{12}c_2 - 90s_1c_1^{11}c_2^2 - 49084s_2c_1^{12}c_2^{13} \\
& + 34758s_2c_1^{10}c_2^{13} + 1960s_2c_1^4c_2^{13} - 3214s_2c_1^8c_2^{13} - 5512s_2c_1^6c_2^{13} \\
& - 186s_2c_1^2c_2^{13} - 32s_1s_2c_1^3c_2^{11} - 288s_1s_2c_1^3c_2^{13} - 5720s_1s_2c_1^5c_2^{11} \\
& + 5288s_1s_2c_1^5c_2^{13} + 44672s_1s_2c_1^7c_2^{11} - 28816s_1s_2c_1^7c_2^{13} \\
& - 132516s_1s_2c_1^9c_2^{11} + 72372s_1s_2c_1^9c_2^{13} - 132516s_1s_2c_1^{11}c_2^9 \\
& - 84592s_1s_2c_1^{11}c_2^{13} + 72372s_1s_2c_1^{13}c_2^9 - 84592s_1s_2c_1^{13}c_2^{11} \\
& - 28816s_1s_2c_1^{13}c_2^7 - 32s_1s_2c_1^{11}c_2^3 - 288s_1s_2c_1^{13}c_2^3, \quad (6.52)
\end{aligned}
$$

$$
\begin{aligned}
n_R^{(4)C} :=& -5720s_1s_2c_1^{11}c_2^5 + 44672s_1s_2c_1^{11}c_2^7 + 5288s_1s_2c_1^{13}c_2^5 \\
& + 12s_1s_2c_1^{11}c_2 - 12s_1s_2c_1^{13}c_2 - 12s_1s_2c_1c_2^{13} + 12s_1s_2c_1c_2^{11} \\
& + 14504c_1^6c_2^{14} - 1552c_1^4c_2^{14} - 26392c_2^{12}c_1^6 + 122669c_2^{12}c_1^8
\end{aligned}
$$

$$- 264682c_2^{12}c_1^{10} + 269177c_2^{12}c_1^{12} - 102868c_2^{12}c_1^{14} - 26392c_1^{12}c_2^6$$
$$+ 122669c_1^{12}c_2^8 - 264682c_1^{12}c_2^{10} - 1552c_1^{14}c_2^4 + 14504c_1^{14}c_2^6$$
$$- 56702c_1^{14}c_2^8 + 110054c_1^{14}c_2^{10} - 2s_1c_1^{13} + 6c_1^{14}c_2^2 + 6c_1^2c_2^{14}$$
$$+ 70c_1^2c_2^{12} + 1771c_1^{12}c_2^4 + 70c_1^{12}c_2^2 + 1771c_1^4c_2^{12} + 2s_2c_2^{13}$$
$$- 102868c_1^{12}c_2^{14} - 56702c_1^8c_2^{14} + 110054c_1^{10}c_2^{14} + 36560s_1s_2c_1^{13}c_2^{13}$$
$$- 21276s_1c_1^{13}c_2^{14} + 21276s_2c_1^{14}c_2^{13} + 36556c_1^{14}c_2^{14}. \tag{6.53}$$

Finally, the denominator of the scalar curvature is given by the following trigonometric polynomials:

$$r_R^{(1)C} := - 210c_2^4c_1^6 + 252c_1^5c_2^5s_1s_2 - 68c_1^4c_2^8 + 3660c_1^6c_2^8$$
$$- 68c_1^8c_2^4 + 3660c_1^8c_2^6 - 45c_1^2c_2^8 - 45c_1^8c_2^2$$
$$- 31c_1^2c_2^{10} + 991c_1^4c_2^{10} - 5074c_1^6c_2^{10} + 5257c_1^8c_2^{10}$$
$$+ 5257c_1^{10}c_2^8 - 5074c_1^{10}c_2^6 - 31c_1^{10}c_2^2 + 991c_1^{10}c_2^4$$
$$+ 172c_1^5c_2^7s_1s_2 + 120c_1^7c_2^3s_1s_2 + 120c_1^3c_2^7s_1s_2 + 172c_1^7c_2^5s_1s_2$$
$$+ 10s_1s_2c_1^9c_2 + 10s_1c_2^9s_2c_1 + 5916c_1^7c_2^5s_1s_2 + 138c_1^9c_2^3s_1s_2$$
$$- 2132c_1^9c_2^5s_1s_2 + 5916c_1^9c_2^7s_1s_2 + 138s_1c_2^9s_2c_1^3 - 2132s_1c_2^9s_2c_1^5$$
$$- 10970c_1^8c_2^8 - 210c_1^4c_2^6 - 672s_1c_2c_1^5 + 672s_2c_2^5c_1^6 - c_1^{10}$$
$$- 4112c_1^7c_2^7s_1s_2 + 70c_1^6c_2^6 - c_2^{10} + 6836c_1^{11}c_2^{11}s_1s_2 + 8265c_1^{10}c_2^{10}, \tag{6.54}$$

$$r_R^{(2)C} := - 310c_1^9c_2^9s_1s_2 - 292s_2c_1^4c_2^{11} + 2840s_2c_1^6c_2^{11} - 7846s_2c_1^8c_2^{11}$$
$$+ 8556s_2c_1^{10}c_2^{11} - 3240s_2c_1^{12}c_2^{11} + 8876s_2c_1^8c_2^9 - 20s_2c_1^2c_2^{11}$$
$$- 220s_2c_1^8c_2^7 - 1596s_2c_1^8c_2^5 - 120s_2c_1^{10}c_2^5 + 6560s_2c_1^{10}c_2^7$$
$$- 14604s_2c_1^{10}c_2^9 + 1116s_2c_1^{12}c_2^5 - 4784s_2c_1^{12}c_2^7 + 6942s_2c_1^{12}c_2^9$$
$$- 8876s_1c_1^9c_2^8 + 14604s_1c_1^9c_2^{10} - 6942s_1c_1^9c_2^{12} - 2840s_1c_1^{11}c_2^6$$
$$+ 7846s_1c_1^{11}c_2^8 - 8556s_1c_1^{11}c_2^{10} + 3240s_1c_1^{11}c_2^{12} - 688s_2c_1^6c_2^9$$
$$- 24s_2c_1^{12}c_2^3 + 688s_1c_1^9c_2^6 - 618s_2c_1^4c_2^9 + 292s_1c_1^{11}c_2^4$$
$$+ 220s_1c_1^7c_2^8 - 6560s_1c_1^7c_2^{10} + 618s_1c_1^9c_2^4 + 2080s_1c_1^7c_2^6$$
$$+ 4784s_1c_1^7c_2^{12} + 24s_1c_1^3c_2^{12} - 1116s_1c_1^5c_2^{12} + 120s_1c_1^5c_2^{10}, \tag{6.55}$$

$$r_R^{(3)C} := - 284s_2c_1^{10}c_2^3 + 1596s_1c_1^5c_2^8 - 2080s_2c_1^6c_2^7 + 284s_1c_1^3c_2^{10}$$
$$+ 10s_1c_1c_2^{12} - 10s_2c_1^{12}c_2 + 20s_1c_1^{11}c_2^2 - 258c_1^3c_2^8s_1$$
$$- 20c_1c_2^{10}s_1 + 258c_1^8c_2^3s_2 - 492c_1^7c_2^5s_1 + 492c_1^4c_2^7s_2$$
$$- 196s_1s_2c_1^3c_2^{11} + 748s_1s_2c_1^5c_2^{11} + 768s_1s_2c_1^7c_2^{11} - 6886s_1s_2c_1^9c_2^{11}$$
$$- 6886s_1s_2c_1^{11}c_2^9 - 196s_1s_2c_1^{11}c_2^3 + 748s_1s_2c_1^{11}c_2^5 + 768s_1s_2c_1^{11}c_2^7$$

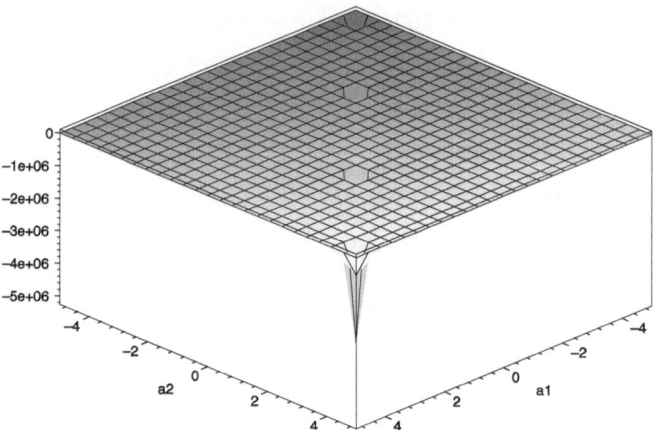

Fig. 6.8 Determinant of the metric tensor plotted as a function of the power factors a_1 and a_2, describing the complex power fluctuations in electrical networks

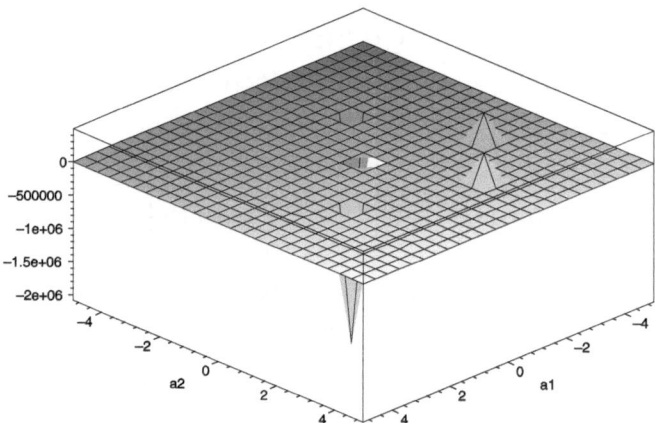

Fig. 6.9 Curvature scalar plotted as a function of the power factors a_1 and a_2, describing the complex power fluctuations in electrical networks

$$
\begin{aligned}
&+ 10s_1s_2c_1^{11}c_2 + 10s_1s_2c_1c_2^{11} + 794c_2^{12}c_1^6 + 3366c_2^{12}c_1^8 \\
&- 10303c_2^{12}c_1^{10} + 6835c_2^{12}c_1^{12} + 794c_1^{12}c_2^6 + 3366c_1^{12}c_2^8 \\
&- 10303c_1^{12}c_2^{10} + 53c_1^2c_2^{12} - 489c_1^{12}c_2^4 + 53c_1^{12}c_2^2 - 489c_1^4c_2^{12} \\
&+ 2c_2^{11}s_2 - 2c_1^{11}s_1 + 92c_1^2c_2^9s_2 + 20c_1^{10}c_2s_2 - 92c_1^9c_2^2s_1.
\end{aligned}
\tag{6.56}
$$

For the choices $V = 1$ and $R_0 = 1$, Fig. 6.8 shows that the determinant of the metric tensor has a cumulative effect on the real and complex power flow fluctuations. This plot elucidates the nature of the stability of a joint power flow in a realistic electrical network. The corresponding plot for the scalar curvature is depicted in Fig. 6.9. This plot shows the global nature of combined real and imaginary power flows

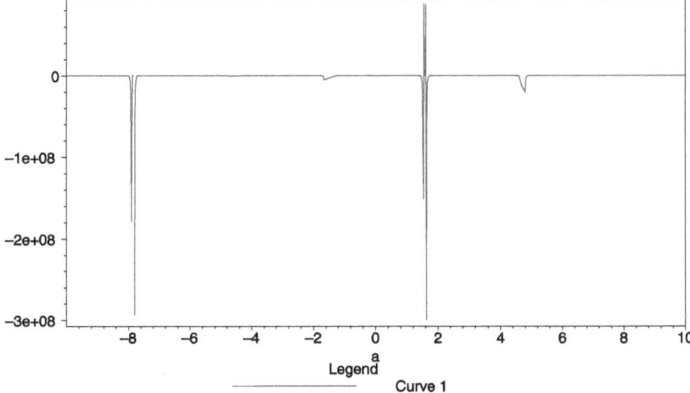

Fig. 6.10 Determinant of the metric tensor plotted as a function of the equal power factors $a :=$ $a_1 = a_2$, describing the complex power fluctuations in electrical networks

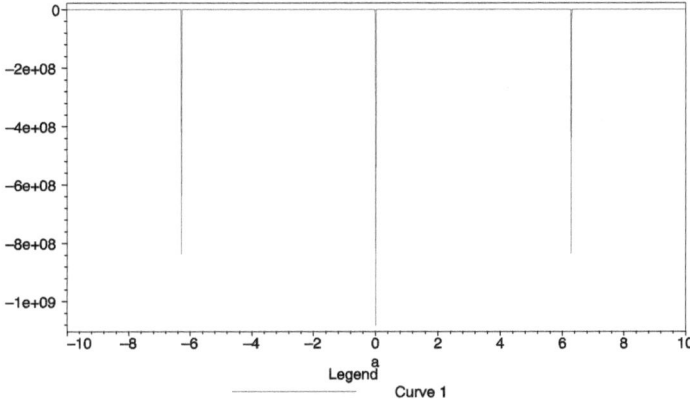

Fig. 6.11 Curvature scalar plotted as a function of the equal power factors $a := a_1 = a_2$, describing the complex fluctuations in electrical networks

in the electrical network. This analysis remains valid under the effect of Gaussian fluctuations of the resistance, reactance, and impedance in a realistic network.

For equal network parameters $a_1 = a$ and $a_2 = a$, the surface plots of the determinant of the metric tensor and scalar curvature are shown in the Figs. 6.10 and 6.11, respectively. We observe here that the stability of the network power flow exists in certain distorted bands. These distortions are strong enough to be able to modulate the global properties of the power flow fluctuations. Specifically, for equal values of the network power factors, global instabilities exist for three specific values.

References

1. G. Radman, R.S. Raje, Power flow model/calculation for power system with multiple FACTS controllers. Elsevier Sci. Dir. Electr. Power Syst. Res. **77**, 1521–1531 (2007)
2. J. Grainger Jr., W. Stevenson, *Power System Analysis*, 1st edn. (McGraw-Hill Science, Engineering, Math, New York 1994)
3. G. Ruppeiner, Riemannian geometry in thermodynamic fluctuation theory. Rev. Mod. Phys. **67**, 605 (1995); [Erratum **68**, 313 (1996)]

Chapter 7
Phase Shift Correction

In the context of network planning, equipment and load dynamics are of course the driving force behind phase shift voltage instability. In this chapter, for the 2-bus systems in a connected power system, we show that the voltage of all relevant buses (i.e., both bus-1 and bus-2) varies in the same way as the transmitted power. In the sequel, we examine the state-space formulation pertaining to voltage regulation and phase-shift correction. In the state-space formulation, some assumptions are made, i.e., shunt admittance has been neglected, there is no reactive support on the load bus, and the generator terminal voltage phasor is assumed to coincide with the rotor position, the load being defined by the real and reactive power demand. We anticipate further that the incorporation of shunt admittance, reactive support on the load bus, and the generator terminal voltage phasor will involve a sub-dominant correction.

From the state-space perspective, large amounts of data concerning power systems are available for analysis, but have not so far been efficiently utilized, due to the lack of efficient methods [1]. Using the notion of state-space geometry, we extend the analysis of power system instability due to parametric fluctuations to deal with voltage regulation and associated phase shift corrections. In practice, this can be utilized to analyze the abundant available data. Highly deregulated power networks generate uncertainties in the operating conditions for given power systems. In particular, we calculate a set of realistic aspects of electrical power flows and their technical details. We also extend the network design for a constant longitudinal phase, as required in order to maintain an economically viable power system from the operating point of view, optimizing the parameter values in such a way as to prevent network blackouts. In particular, we exploit the fact that there exists an equilibrium network configuration.

With the intrinsic geometric characterization as introduced in Chap. 2, we systematically analyze the underlying stability structures of the network in the real and imaginary power flows, together with their joint effects on the network when a finite number of the chosen parameters are allowed to fluctuate. The state-space formulation of the voltage regulation problem enables one to stabilize the

S. Bellucci et al., *Geometrical Methods for Power Network Analysis*,
SpringerBriefs in Electrical and Computer Engineering,
DOI: 10.1007/978-3-642-33344-6_7, © The Author(s) 2013

fluctuations in the maximum deliverable power through a load [2]. In a practical power system, the power flow occurs at some dropping voltage, and our proposal ensures the limiting instability. For a given generator voltage, this analysis of the fluctuations is well suited for determining the allowed operating regimes. For a given recovery voltage, we find that the fluctuations about a sinusoidal steady-state regime can be controlled by a parabolic power flow. The network correction properties offer an understanding of the underlying domain for feasible power system operation.

As far as engineering aspects are concerned, it turns out that the voltage regulation and phase shift correction problem can be extended to deal with a series of complex filters, non-steady regimes, and large voltage disturbances. The detailed computations and plots of the underlying state-space quantities provide a modern way to examine the performance of, and plan for, a deregulated power market. In order to realize the above scenarios, the present methodology depends on the nature of the model parameters. By considering a given set of (input) parameters, we provide a precise and detailed approach to network planning. We extend the state-space analysis to non-steady state regimes and large scale voltage instability.

Notice the fact that the voltage depends not only on the electromotive force of the generator, but also on the behavior of the load, grid capacity, and reactance [3–5]. Hence, the quality of the power system refers not only to the continuity of the supply, but also to the variation of the voltage and frequency within the given operating limit. Apart from the voltage drop occurring due to industrial consumption of reactive power, the network itself contributes to the supply and absorption of reactive power, while the real power is almost solely a function of the voltage phase difference t. Indeed, there have been numerous examinations of phase shift corrections pertaining to the voltage regulation problem (see, for example, [6, 7]). Here, modifications aimed at voltage phase correction are carried out with the help of phase-shifting transformers. The reactive support is required in addition to the phase corrections in order to stabilize the power system. Thus, to ensure the stability of the power system, the present analysis is carried out by taking the complex power instead of only the real power.

With regard to the voltage regulation problem [2], let us consider a stabilization manifold with the fluctuation coordinates $x^a = (t, s) \in \mathcal{M}_2$, where t is the phase difference between bus voltages and s is the phase shift by the transformer to control the stability of the power systems. Thus, the line element of voltage fluctuation is given by

$$ds^2 = \frac{\partial^2}{\partial t^2} S(t, s) dt^2 + 2 \frac{\partial^2}{\partial t \partial s} dt \, ds + \frac{\partial^2}{\partial s^2} S(t, s) ds^2, \qquad (7.1)$$

where S is the complex power defined by $S = P + JQ$, with P and Q the real and reactive power supplied to the load and J the imaginary unit. In fact, a practical power system is highly non-linear in nature, and thus the present consideration of fluctuations gives rise to a new framework for investigating the stability of the system.

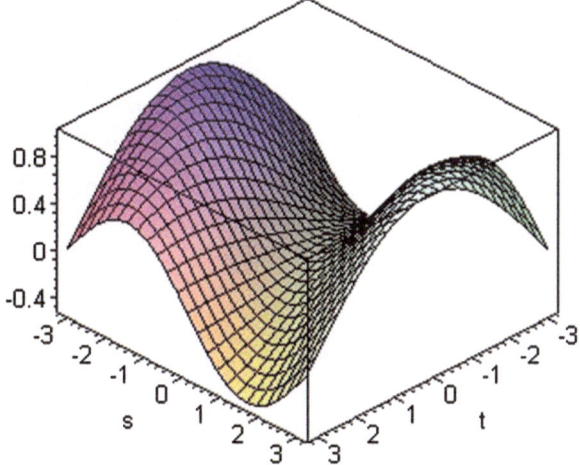

Fig. 7.1 The tt component of the metric tensor plotted as a function of $\{t, s\}$, describing fluctuations in the sinusoidal flow of the complex power network

Under a general consideration of sinusoidal flow, it is shown in [2] that the complex power is

$$S(t, s) = -4\frac{V^2}{X} \sin\left(-\frac{1}{2}t + \frac{1}{2}s\right) \sin\left(\frac{1}{2}s\right). \tag{7.2}$$

For a given voltage level, the limit of the real and the reactive power can be defined for any apparatus as in (7.2). Here, X can be defined as the aggregated form of the load and the line impedance, for simplicity of formulation. This can be extended by separating the line and load impedances. For the required complex power flow, as discussed in the context of state-space geometry, we find that the components of the fluctuations are

$$g_{tt} = \frac{V^2}{X} \sin\left(-\frac{1}{2}t + \frac{1}{2}s\right) \sin\left(\frac{1}{2}s\right), \tag{7.3}$$

$$g_{ts} = -\frac{V^2}{X}\left[\sin\left(-\frac{1}{2}t + \frac{1}{2}s\right) \sin\left(\frac{1}{2}s\right) - \cos\left(-\frac{1}{2}t + \frac{1}{2}s\right) \cos\left(\frac{1}{2}s\right)\right], \tag{7.4}$$

$$g_{ss} = 2\frac{V^2}{X}\left[\sin\left(-\frac{1}{2}t + \frac{1}{2}s\right) \sin\left(\frac{1}{2}s\right) - \cos\left(-\frac{1}{2}t + \frac{1}{2}s\right) \cos\left(\frac{1}{2}s\right)\right]. \tag{7.5}$$

At a given S, in order to analyze the instability occurring due to a voltage regulation, for the input voltage of $V = 1$ p.u. and initial load $X = 1$ p.u., Figs. 7.1 and 7.2 show the fluctuations in the diagonal components $\{g_{tt}, g_{ss}\}$ of the metric tensor. The value of X depends on the design and the load of the network, and so can vary from system to system, but the procedure of the state-space analysis would be the same.

In the regime $t \in (-3, 3)$ and $s \in (-3, 3)$, we notice that the amplitude of $\{g_{tt}\}$ takes a value between -0.5 and $+1$. In this range of the parameters $\{t, s\}$,

Fig. 7.2 The ss component
of the metric tensor plotted as
a function of $\{t, s\}$, describing
fluctuations in the sinusoidal
flow of the complex power
network

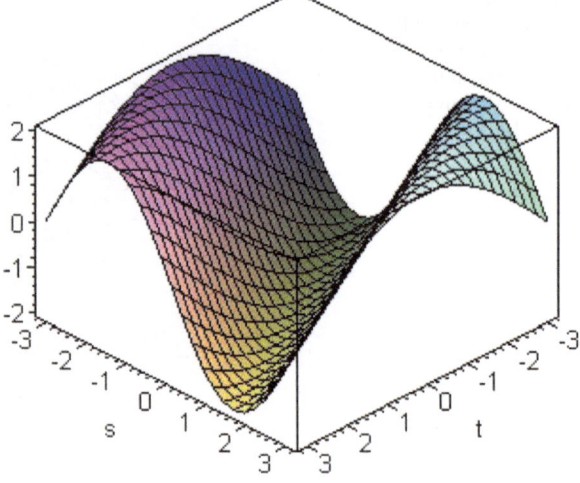

Fig. 7.3 The ts component
of the metric tensor plotted as
a function of $\{t, s\}$, describing
fluctuations in the sinusoidal
flow of the complex power
network

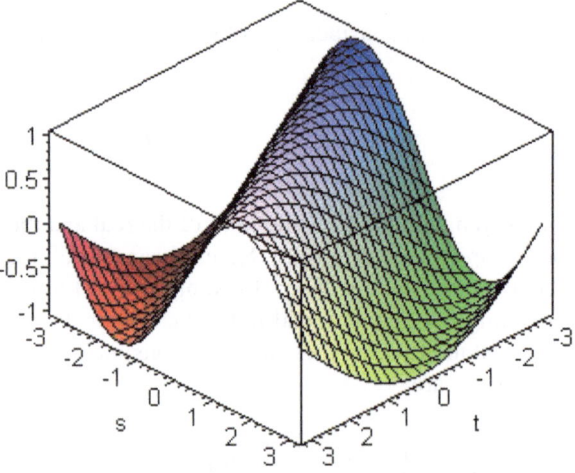

we find that the mix component $\{g_{ts}\}$ lies in the range $(-1, +1)$. In this case, we
see that the range of growth of the amplitude of $\{g_{tt}\}$ remains almost double in the
same limit of the parameters $\{t, s\}$. Explicitly, this signifies that sinusoidal voltage-
regulated electrical networks are thermodynamically unstable in the limit of large s
and small t. Thus, higher order phase corrections are required for large s in order to
stabilize the system, and these can easily be extracted from the components of the
state-space metric tensor.

Similarly, Fig. 7.3 shows the behavior of the $\{g_{ts}\}$ component of the state-space
metric tensor. We find that the mix component $\{g_{ts}\}$ takes a uniform value of order
± 1 in both limits of the parameters $\{t, s\}$. In this limit of $\{t, s\}$, the local fluctuation
under the voltage regulation problem as depicted in Figs. 7.1, 7.2, and 7.3 illustrates

Fig. 7.4 Determinant of
metric tensor plotted as a
function of $\{t, s\}$, describing
fluctuations in the sinusoidal
flow of the complex power
network

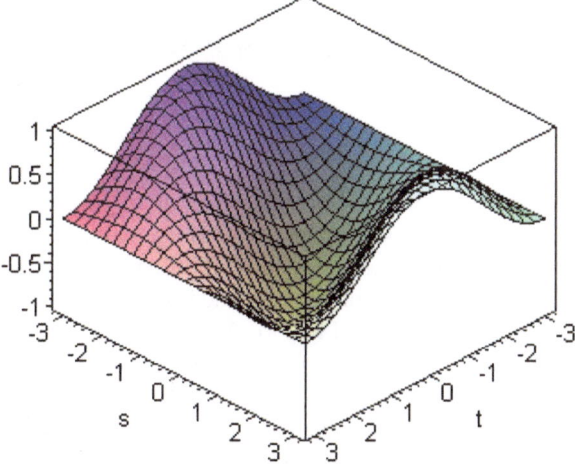

the stability properties of voltage-regulated networks. In short, the self pair fluctuations involving $\{t, s\}$, as defined by the metric tensor $\{g_{ij} | i, j = t, s\}$, have both positive and negative numerical values, so the sinusoidal voltage-regulated networks are stable in a particular domain of the parameters.

For a given operating point, if the diagonal components are negative, then we may choose the phase correction so that it becomes positive, since this negativity signals instability of the power system. Further justification can be made for the global stability by requiring positivity of the determinant of the metric tensor. It is not difficult to show that the determinant of the metric tensor is

$$g(s, t) = -\frac{V^4}{X^2}\left[-1 + \cos^2\left(\frac{1}{2}s\right) + \cos^2\left(-\frac{1}{2}t + \frac{1}{2}s\right)\right]. \qquad (7.6)$$

Notice that the overall stability of voltage-regulated networks can be determined in terms of the values of the regulation parameters t, s. This follows from the behavior of the determinant of the metric tensor. In this case, we find that the determinant of the metric tensor tends to a negative value when the regulation parameters take relatively large absolute values. For $t \in (-3, 3)$ and $s \in (-3, 3)$, Fig. 7.4 shows that the determinant of the metric tensor lies in the interval $(-1, 1)$. In fact, we find that the negativity of g increases as the values of t, s are increased from zero to ± 3. In such cases, the surface defined by the fluctuations of $\{t, s\}$ is unstable. When only one of the parameters is allowed to vary, the stability of the voltage-regulated network configuration is determined by the positivity of the first principal minor. In other words, this amounts to requiring the positivity of the tt component of the metric tensor. The above graphical properties and positivity of the state-space quantities provide the underlying stability properties of the two-parameter voltage regulation network problem.

Interestingly, we observe that the surface of voltage regulation is globally non-interacting for vanishing state-space scalar curvature. Indeed, we have the following Ricci scalar

$$R\,(s,t) = 0, \quad \forall\,(s,t) \in \mathcal{M}_2. \tag{7.7}$$

The global stability properties of voltage-regulated networks follow from the underlying state-space scalar curvature, as we find that the scalar curvature vanishes identically for all values of the regulation parameters. In this case, this shows that the fluctuating output voltage-regulated networks correspond to a noninteracting statistical configuration. In short, the above observations of the state-space geometry indicate that, although voltage-regulated networks are noninteracting in the global sense, they nevertheless correspond to stably regulated network configurations in a specific domain of the regulation parameters. In particular, when the parameters $\{t, s\}$ are allowed to fluctuate, there exists a certain domain of the regulation parameters in which some of the components can fail.

The above procedure delivers the phase correction values s for a specific set of fluctuations in t for which the determinant of the metric tensor goes beyond the limits of the given complex power. This observation follows from the fact that there are non-trivial instabilities at the local level of the parametric fluctuations. From the above observation, it can be seen that the iterative procedure of selecting the regulation parameters can be replaced by a non-iterative direct method derived from the state-space geometry. This formulation incorporates fluctuations of the parameters which follow the non-linearity of the system. It is worth mentioning here that this method can even be applied to a large network with minimal computational effort.

References

1. F.F. Wu, Power system state estimation: a survey. Int. J. Electr. Power Energy Syst. **12**(2), 80–87 (1990)
2. S.K. Biswas, N.U. Ahmed, Optimal voltage regulation of power systems under transient conditions. Electr. Power Syst. Res. **6**(1), 71–77 (1983)
3. M. Crappe, *Electric Power Systems* (ISTE Ltd/Wiley, New York, 2008)
4. A.R. Bergen, V. Vittal, *Power System Analysis*, 2nd edn. (Prentice Hall, Upper Saddle River, 2000)
5. J.D. Glover, M.S. Sarma, *Power Systems Analysis and Design*, 3rd edn. (Brooks/Cole, USA, 2002)
6. M. Gremia, J. Trecat, A. Germond, *Réseaux électriques, aspects actuels* (Editura Technica, Bucharest, 2000)
7. M. Auet, J.J. Mort, *Energie électrique* (Georgio, Hausanne, 1981)

Chapter 8
Complex Power Optimization

In this chapter, we analyze the voltage instability pertaining to the maximum deliverable power for a given load. We thus illustrate the role of state-space geometry in complex power flow optimization. In the rapidly growing and competitive power market, optimization is crucial in order to define the loadability limit of the power network, where not only the real power is considered, but the reactive support is also involved. Notice that, for complex power optimization, the inequality constraint is

$$\sum_{j=1}^{p} s_{ij} \Delta_j - c_j \leq 0. \tag{8.1}$$

The above inequality is the criterion applied to prevent the system from a potentially dangerous incident. It is used to determine preventive controls by means of a set of economic and technical safeguards. Δ_j is the variation of the j th control variable and the coefficient s_{ij} can be determined by eigenvalue analysis (see, for instance, [1]). In the corresponding series filter, we have the impedance

$$X = \sqrt{X_L^2 + R^2 + X_C^2}, \tag{8.2}$$

where

$$X_L = wL, \quad X_C = \frac{1}{wC}, \tag{8.3}$$

with R the resistance, L the reactance, and C the series capacitance of the transmission line. In fact, this formulation can be extended to cover the case of a nonzero shunt capacitance, provided the load bus is also taken into account in X_C. In the steady-state sinusoidal regime, the power flow equations of the above system are

$$P = -\frac{EV}{X_L} \sin t, \quad Q = -\frac{V^2}{X} + \frac{EV}{X} \cos t. \tag{8.4}$$

S. Bellucci et al., *Geometrical Methods for Power Network Analysis*,
SpringerBriefs in Electrical and Computer Engineering,
DOI: 10.1007/978-3-642-33344-6_8, © The Author(s) 2013

Eliminating t from these power flow equations, we see in this case that the output voltage can be expressed as

$$V(P, Q) = \sqrt{\frac{E^2}{2} - QX \pm \sqrt{\frac{E^4}{4} - X^2 P^2 - X E^2 Q}}. \tag{8.5}$$

We now introduce scaling $V \to V/E$ and new variables

$$x = \frac{QX}{E^2}, \quad y = \frac{PX}{E^2}. \tag{8.6}$$

The values of x and y can be determined by the varying values of Q, P, and X. Thus, it follows that the complex power corrected voltage can be written as

$$V(x, y) := \sqrt{\frac{1}{2} - x + k\sqrt{\frac{1}{4} - y^2} - x}, \tag{8.7}$$

where the parameter k is the sign function, taking values in the set $\{+1, -1\}$. Generally, $+1$ is acceptable for power system instability analysis, but we shall nevertheless present our study for both signatures. The analysis for the two values of k defines the limits of the complex power in order to keep the power system stable. To illustrate the effectiveness of the results, we carry out the analysis on two variables. In a large network, more variables can be included in order to specify the operating limit.

In order to illustrate the power of state-space geometry, let us consider a two-dimensional vector $x^a = (x, y)$ spanning an inner product space like the two-dimensional manifold (M_2, g) with line element

$$ds^2 = \frac{\partial^2 V(x, y)}{\partial x^2} dx^2 + 2\frac{\partial^2 V(x, y)}{\partial x \partial y} dx\, dy + \frac{\partial^2 V(x, y)}{\partial y^2} dy^2. \tag{8.8}$$

The above analysis is carried out with respect to the voltage and we may easily check whether the specified voltage is stable or not. For a given complex power, we will express this criterion in terms of the elements of the metric tensor. Here, we find that the components of the metric tensor reduce to the following expressions:

$$g_{xx} = \frac{4zy^2 + 4zx - 4k + 8ky^2 + 12kx - 4z}{(-1 + 2x - kz)\sqrt{2 - 4x + 2kz}(-1 + 4y^2 + 4x)z}, \tag{8.9}$$

$$g_{xy} = \frac{2ky(-3 + 4y^2 + 8x - 3kz)}{(-1 + 2x - kz)\sqrt{2 - 4x + 2kz}(-1 + 4y^2 + 4x)z}, \tag{8.10}$$

$$g_{yy} = -2k\frac{8x^2 - 6x - 4kzx + 1 + kz + 2y^2kz}{z(-1 + 4y^2 + 4x)\sqrt{2 - 4x + 2kz}(-1 + 2x - kz)}. \tag{8.11}$$

Fig. 8.1 The xx component of the metric tensor plotted as a function of $\{x, y\}$, describing the fluctuations in complex power corrected voltage networks for $k = 1$

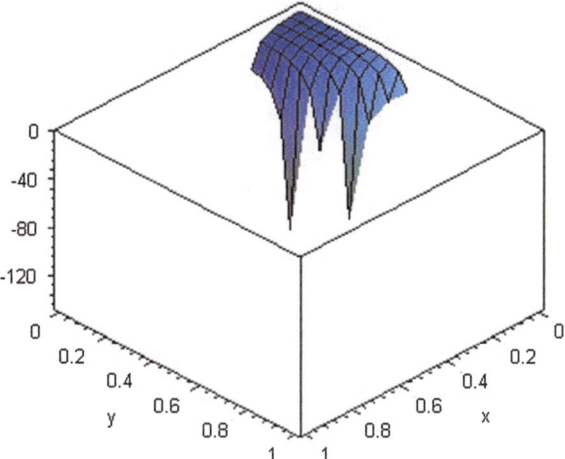

Equations (8.9)–(8.11) are highly non-linear. The fluctuations in each element of the metric tensor thus define the operating point very accurately if we introduce the function $z(x, y)$ defined by

$$z := \sqrt{1 - 4y^2 - 4x}. \tag{8.12}$$

The stability characteristic of a complex power network follows from the positivity of the heat capacities $\{g_{xx}, g_{yy}\}$ of the state-space metric tensor. These are basically the diagonal components of the metric tensor associated with complex power optimization. For the choice $k = 1$, the explicit graphical view of the above-mentioned local fluctuations is depicted in Figs. 8.1 and 8.2.

For the regime $x, y \in (0, 1)$, we see that the amplitude of g_{xx} takes a value of the order of -160. In this range of the parameters $\{x, y\}$, we find that the component g_{yy} lies in the range $(0, -60)$. In this case, we observe that the growth of the amplitude of g_{xx} happens in a rather distinct limit of $\{x, y\}$. In the regime where x takes a given value up to 0.15, the system is stable over the range $y \in (0, 0.4)$. For values of y outside this range, the system tends towards an instability. The voltage for large y leads to an instability and can thus cause a system outage.

As can be seen from (8.4)–(8.6), the power flow of the heat capacities depends on the network parameters, and thus changing the value of a capacity or the fluctuation in X can affect the power deliverable to the load. The diagonal component of the metric tensor should therefore be positive, in which case fluctuations provide a set of values from which one could determine the required local values of the flow parameters x and y. This signifies that complex power optimization could yield a certain thermodynamic instability. One must therefore choose a specific domain of the parameters $\{x, y\}$ such that the desired network remains in a well balanced limit of the power flow parameters.

Fig. 8.2 The yy component of the metric tensor plotted as a function of $\{x, y\}$, describing the fluctuations in complex power corrected voltage networks for $k = 1$

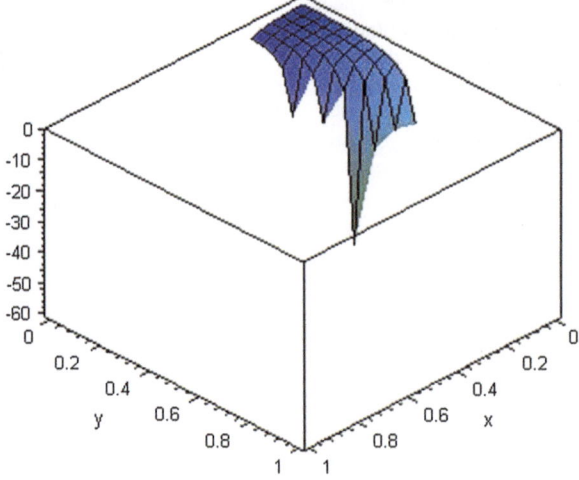

Fig. 8.3 The xy component of the metric tensor plotted as a function of $\{x, y\}$, describing the fluctuations in complex power corrected voltage networks for $k = 1$

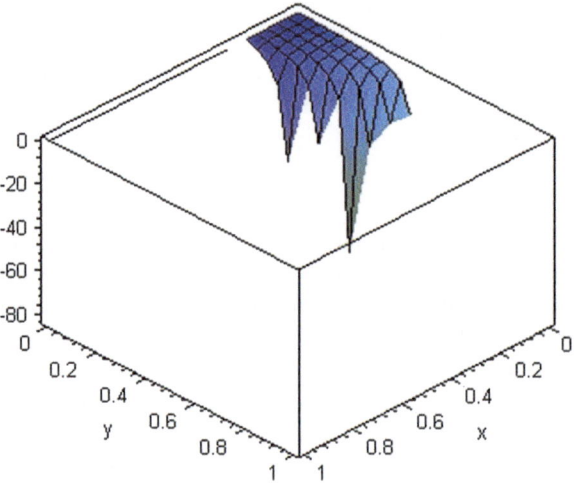

Further, we notice from Fig. 8.3 that the mix component g_{xy} of the state-space metric tensor has a negative value under power flow fluctuations. Interestingly, we find that all the local fluctuations happen in a small limit of the power flow parameters $\{x, y\}$, where higher values of y may cause an instability. Figure 8.3 shows that a large real power cannot flow through the network while reactive power is available. Thus, the other parameters have to be tuned in order to increase the stability limit of the network. For this reason, the network would be planned by increasing the capacity of the transmission lines. In this limit, we see that Figs. 8.1, 8.2, and 8.3 illustrate the local fluctuation properties of the power network under the complex power flow. In fact, both the self pair fluctuations involving $\{x, y\}$, as defined by the

Fig. 8.4 The xx component of the metric tensor plotted as a function of $\{x, y\}$, describing the fluctuations in complex power corrected voltage networks for $k = -1$

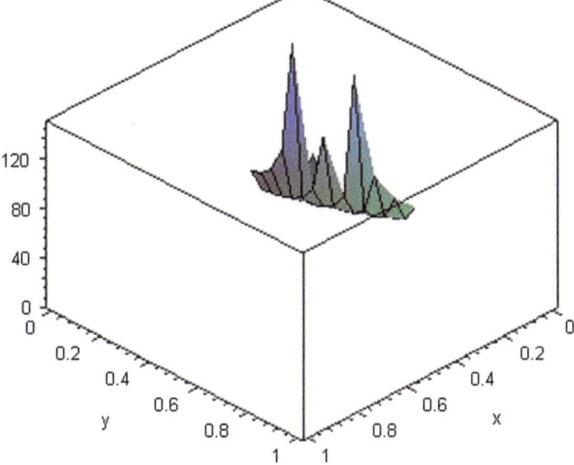

Fig. 8.5 The yy component of the metric tensor plotted as a function of $\{x, y\}$, describing the fluctuations in complex power corrected voltage networks for $k = -1$

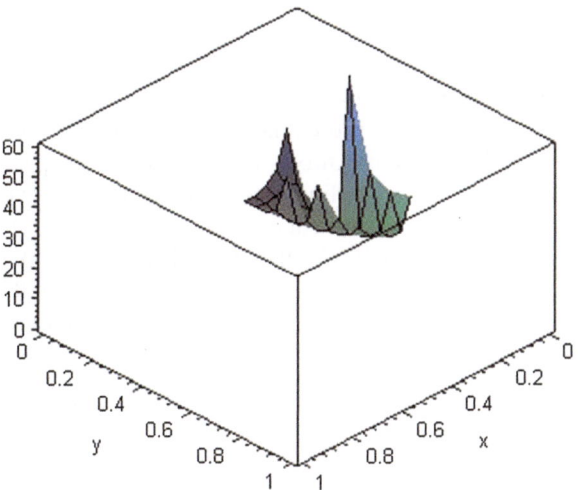

metric tensor $\{g_{ij} | i, j = x, y\}$ and the mix component g_{MQ}, assume only negative numerical values. More precisely, the mix component may be negative, so in order to see the global stability limit, we require the determinant of the metric tensor to be positive. Thus, with the values for a given set of fluctuating $\{x, y\}$ as shown in Figs. 8.1, 8.2, and 8.3, the determinant is illustrated in Fig. 8.7 for $k = 1$. The optimized value(s) of $\{x, y\}$ in the positive determinant regime will keep the power system voltage globally stable.

The heat capacities for the case of $k = -1$ are shown in Figs. 8.4 and 8.5. In this case, in the interval $x, y \in (0, 1)$, the amplitude of g_{xx} takes a positive value of the order of 160. In this range of the parameters $\{x, y\}$, Fig. 8.5 shows that the

Fig. 8.6 The xy component of the metric tensor plotted as a function of $\{x, y\}$, describing the fluctuations in complex power corrected voltage networks for $k = -1$

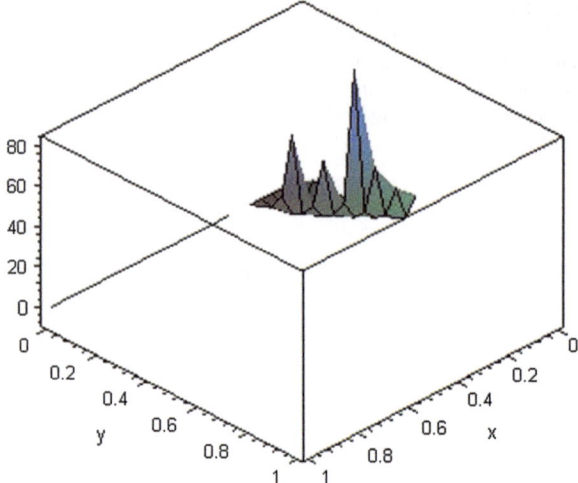

component g_{yy} lies in the range $(0, 60)$ and we find that the fluctuations of both g_{xx} and g_{yy} maintain a positive amplitude. As mentioned for the case $k = 1$, fluctuations are also generically present near the origin of the flow parameters. Figure 8.6 shows that the corresponding mix component of the complex power flow fluctuation metric takes a maximum amplitude of about 80. The above plot may change for a different network and for a different operating condition, as shown by (8.6). The values of x and y are sensitive to X when E tends to unity in p.u. units. This analysis is easily extended to different sizes of the power network and the complex power flow.

We have the following formula for the determinant of the metric tensor:

$$g(x, y) = k \frac{g_1 + k\tilde{g}_1}{g_2^3 g_3^2 z},$$
(8.13)

where the factors g_1, $\tilde{g}_1$, g_2, and g_3 are defined by

$$g_1 = 32x^3 + 32y^2x^2 - 32x^2 - 72x^2 - 24y^2x - 104y^2x$$
$$+ 46x + 10x - 32y^4 + 36y^2 - 1 + 4y^2 - 7,$$
(8.14)
$$\tilde{g}_1 = -40zx^2 + 28zx + 12zx - 32y^2xz - 3z - 5z + 12y^2z + 12y^2z,$$
(8.15)
$$g_2 = -1 + 2x - kz, \qquad g_3 = -1 + 4y^2 + 4x.$$
(8.16)

As in the case of fluctuating voltage regulation mentioned in the previous chapter, the ensemble stability of a network under the complex power flow fluctuations can be determined in terms of the values of the flow parameters $\{x, y\}$, as defined above. Furthermore, this follows from the behavior of the determinant of the fluctuation metric tensor. In both cases $k = \pm 1$, we observe that the determinant of the metric tensor tends to a positive value. For $k = 1$, we see from Fig. 8.7 that the peak of the

Fig. 8.7 Determinant of the metric tensor plotted as a function of $\{x, y\}$, describing the fluctuations in complex power corrected voltage networks for $k = 1$

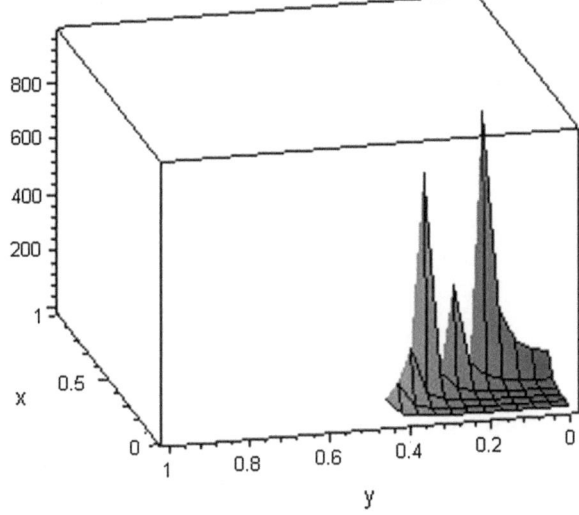

Fig. 8.8 Determinant of the metric tensor plotted as a function of $\{x, y\}$, describing the fluctuations in complex power corrected voltage networks for $k = -1$

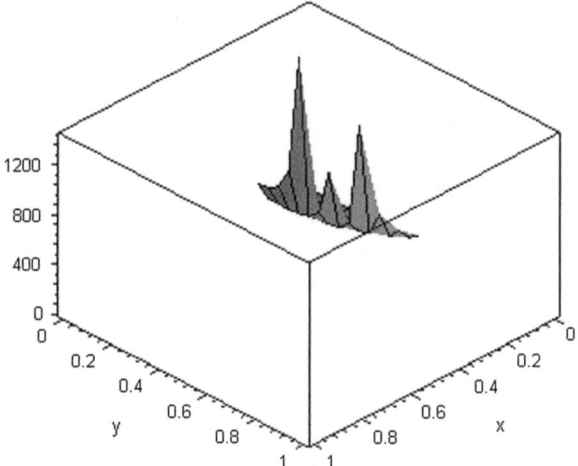

determinant of the metric tensor is of the order 1000, while for $k = -1$, Fig. 8.8 shows that the peak of the determinant of the metric tensor lies in the interval $(0, 1600)$. As in the case of local fluctuations, we find that the global power flow fluctuations also occur for some small values of the flow parameters $\{x, y\}$. When only the parameter x is allowed to vary, the stability of the complex power flow networks is determined by the positivity of the first principal minor $p_1 := g_{xx}$. Physically, the above qualitative demonstrations of the fluctuations illustrate stability properties of two-parameter complex power flow networks.

In general, we find that the long-range correlation can be characterized by the scalar curvature

Fig. 8.9 Curvature scalar
plotted as a function of $\{x, y\}$,
describing the fluctuations
in complex power corrected
voltage networks for $k = 1$

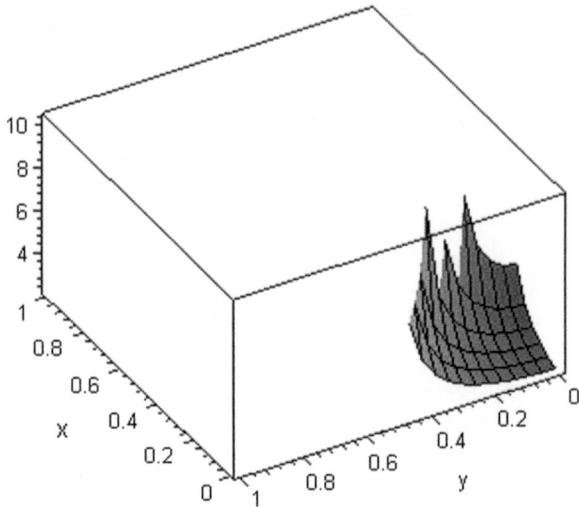

$$R(x, y) = 4\frac{r_1 + k\tilde{r}_1}{(r_2 + k\tilde{r}_2)^3 r_3 z}, \tag{8.17}$$

where the factors r_1, $\tilde{r}_1$, r_2, $\tilde{r}_2$, and r_3 are given as follows. First of all,

$$r_1 = r_{10} + r_{11} + r_{12} + r_{13} + r_{14} + r_{15}, \quad \tilde{r}_1 = \tilde{r}_{11} + \tilde{r}_{12} + \tilde{r}_{13} + \tilde{r}_{14} + \tilde{r}_{15}. \tag{8.18}$$

In this case, it turns out that the specific factors r_{10}, r_{11}, r_{12}, r_{13}, r_{14}, and r_{15} are
as given in Appendix A. Similarly, the factors with the tilde terms are given in
Appendix B. Finally, the factors in the denominator of the scalar curvature are

$$r_2 = 32x^3 - 32x^2 - 72x^2 + 32y^2x^2 - 104y^2x + 46x$$
$$+ 10x - 24y^2x - 32y^4 + 36y^2 - 1 + 4y^2 - 7, \tag{8.19}$$
$$\tilde{r}_2 = -40zx^2 + 28zx + 12zx - 32y^2xz - 3z - 5z + 12y^2z + 12y^2z, \tag{8.20}$$
$$r_3 = \sqrt{2 - 4x + 2kz}. \tag{8.21}$$

The global stability properties of complex power flow networks follow from the
corresponding state-space scalar curvature. In particular, for $k = 1$ and in the range
$x, y \in (0, 1)$, Fig. 8.9 shows that the scalar curvature has a positive amplitude of
order 10. This shows that the underlying network configuration is an interacting
system. The positive sign of the scalar curvature signals the repulsive nature of
interactions.

Figure 8.10 shows the behavior of the above scalar curvature for the same range
of the parameters but for the value $k = -1$. For the case $x \in (0, 1)$ and $y \in (0, 1)$,
we notice from Fig. 8.10 that there is a large negative peak of the global interactions

Fig. 8.10 Curvature scalar
plotted as a function of $\{x, y\}$,
describing the fluctuations
in complex power corrected
voltage networks for $k = -1$

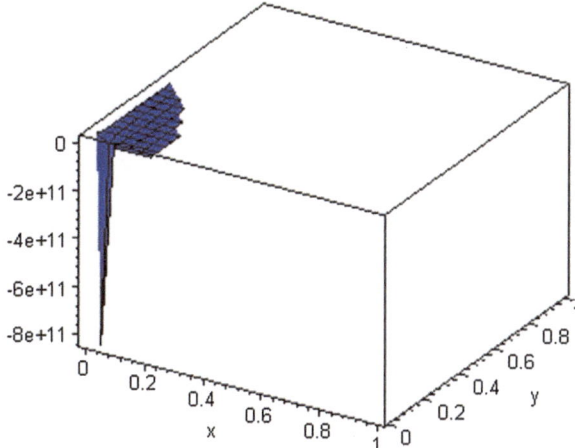

of the order $-8 \times 10^{+11}$. In comparison with the interactions appearing for $k = 1$,
the amplitude of the global interactions appears to be much larger for $k = -1$. Up
to a small range of x and y, the two-parameter power system behaves as a stable
configuration. However, increasing the value of y cannot increase the limit of x, even
though it makes the determinant negative, and thus decreases the real power flow
capacity of the network.

In short, when the parameters $\{x, y\}$ are allowed to fluctuate, the above figures
indicate that complex power flow networks correspond to unstable interacting configurations. The above instability analysis illustrates the operation of a power system
in both domains defined by $k = \pm 1$, for a low voltage profile and a high voltage
profile of the given network parameters. Sensitivity analysis of the network parameters can be used by planners to determine the optimal value of the transmission line
parameters, e.g., the capacity and length of the line.

Reference

1. M. Crappe, *Electric Power Systems* (ISTE Ltd, New York, 2008)

Chapter 9
Large Scale Voltage Instability

In this chapter, we apply the methods of state-space geometry to the issue of complex power optimization in a non-steady state regime by incorporating the intrinsic state-space geometry. Considering the power flow between two arbitrary points along the chosen transmission line, we compute the stability domains for a large voltage instability problem. In the proposed formulation of this problem, some assumptions are made, i.e., shunt admittance has been neglected, we assume that there is no reactive support on the load bus, the generator terminal voltage phasor is assumed to coincide with the rotor position, and the load is defined by the real and reactive power demand.

The voltage instability results from the operating point lying beyond the limit of maximum deliverable power. For a given nonzero $X := (L, C, R)$, when there are fluctuations in the input real and imaginary power, there would be fluctuations in the voltage. In a steady-state configuration, there are a number of important networks, namely the VAN networks. In the present analysis, we have used an aggregated load model by considering the complex power flows defined separately by the following polynomial laws for the real and reactive powers:

$$P = P_0 \left(\frac{V}{V_0}\right)^a, \qquad Q = Q_0 \left(\frac{V}{V_0}\right)^b, \tag{9.1}$$

where V_0 is the initial/reference voltage at the load bus, and P_0 and Q_0 are the real and reactive powers consumed at the initial/reference operating voltage. The reactive support at the load bus can be provided by a shunt capacitor. The above aggregated load model approximates the effect of sub-transmission and distribution lines. Hence, the net power flow defined as the absolute value of $S = P + JQ$ is

$$|S| = \sqrt{P^2 + Q^2} = \sqrt{P_0^2 \left(\frac{V}{V_0}\right)^{2a} + Q_0^2 \left(\frac{V}{V_0}\right)^{2b}}, \tag{9.2}$$

S. Bellucci et al., *Geometrical Methods for Power Network Analysis*, 77
SpringerBriefs in Electrical and Computer Engineering,
DOI: 10.1007/978-3-642-33344-6_9, © The Author(s) 2013

where P_0, Q_0, and V_0 are real constants. The formulation of (9.2) shows that the reactive power affects not only the voltage, but also the real power, and thus the complex power. However, the whole analysis is based on the static load model, which is represented as a function of the voltage. In order to exploit the voltage instability problem, we may consider the output voltage V and real flow exponents $\{a, b\}$ ranging from 0 to 2 as representing a constant power, constant current, and constant impedance load, respectively. We thus let V_0 be the input voltage and (P_0, Q_0) a given input complex power, whence the total output complex power can be expressed as

$$S(a, b, V) := \sqrt{P_0^2 \left(\frac{V}{V_0}\right)^{2a} + Q_0^2 \left(\frac{V}{V_0}\right)^{2b}}. \tag{9.3}$$

In this case, given P_0, Q_0, and V_0, it follows that the manifold of the complex power fluctuation is characterized by the network parameters as coordinates $x^i = (a, b, V) \in M_3$. Indeed, it is not difficult to see that the underlying line element takes the form

$$ds^2 = \frac{\partial^2 S(a, b, V)}{\partial a^2} da^2 + 2\frac{\partial^2 S(a, b, V)}{\partial b \partial a} da\, db + 2\frac{\partial^2 S(a, b, V)}{\partial a \partial V} da\, dV$$
$$+ \frac{\partial^2 S(a, b, V)}{\partial b^2} db^2 + 2\frac{\partial^2 S(a, b, V)}{\partial b \partial V} db\, dV + \frac{\partial^2 S(a, b, V)}{\partial V^2} dV^2. \tag{9.4}$$

The expression for the components of fluctuations are given in Appendix C.

For $P_0 = 1$ p.u., $Q_0 = 1$ p.u., and an arbitrary uniform index t satisfying $a = t = b$, and in the range $t \in (0, 1)$ and $V \in (0, 1)$, we find that the fluctuation of the components g_{aa} and g_{bb} lies in the interval $(0, 10)$, while the component g_{VV} typically takes a value in the interval $(-70, 0)$. This shows that large scale networks are locally unstable configurations with respect to output voltage fluctuations. In fact, the range of growth of g_{aa}, g_{bb}, and g_{VV} happens to occur in a distinct limit of the flow parameters $\{a, b, V\}$.

Explicitly, Figs. 9.1 and 9.2 show that the growth of g_{aa} and g_{bb} takes place in the limit of small t and small V. On the other hand, Fig. 9.3 shows that the growth of the voltage component g_{VV} takes place in the limit of small V and 20% of the total index t. Similarly, the components involving two distinct parameters of the network have been depicted in Figs. 9.4, 9.5, and 9.6. In this case, we notice that Fig. 9.4 shows that the aV component of the fluctuations lies in the interval $(-4, 0)$. However, Figs. 9.5 and 9.6 show that the voltage index components have a positive amplitude of about 16. For a given large scale voltage network, the components of the metric tensor $\{g_{ij}|i, j = a, b, V\}$ indicate that the fluctuations involving the flow indexes $\{a, b\}$ have more positive numerical values than those involving the output voltage.

In this case, the underlying state-space manifold describing the fluctuations of a large scale voltage instability is of dimension three. Thus, as a function of the

Fig. 9.1 The *aa* component
of the metric tensor plotted as
a function of {*t*, *V*}, describing
the fluctuations in the large
scale voltage instability of
complex power networks

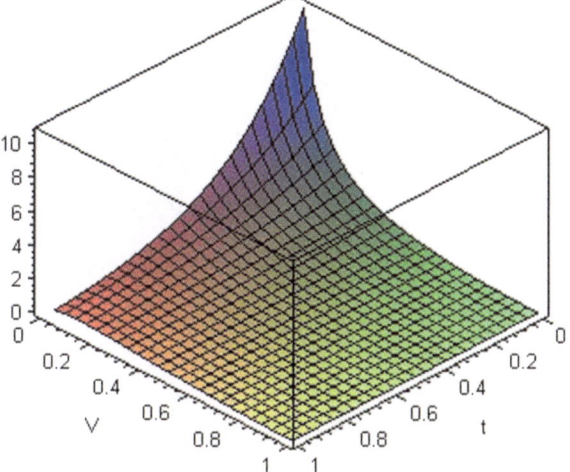

Fig. 9.2 The *bb* component
of the metric tensor plotted as
a function of {*t*, *V*}, describing
the fluctuations in the large
scale voltage instability of
complex power networks

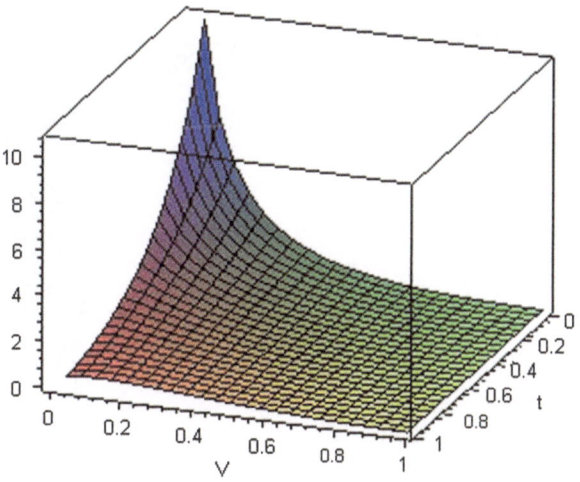

parameters {*a*, *b*, *V*}, it follows easily that the stability of the *ab* surface is characterized by the surface minor

$$p_S := 2 \frac{P_0^2 \ln\left(\frac{V}{V_0}\right)^4 Q_0^2 \left[2\left(\frac{V}{V_0}\right)^{4b+4a} Q_0^2 P_0^2 + \left(\frac{V}{V_0}\right)^{6b+2a} Q_0^4 + \left(\frac{V}{V_0}\right)^{6a+2b} P_0^4 \right]}{\left[P_0^2 \left(\frac{V}{V_0}\right)^{2a} + Q_0^2 \left(\frac{V}{V_0}\right)^{2b} \right]^3}.$$

$$(9.5)$$

Fig. 9.3 The VV component of the metric tensor plotted as a function of $\{t, V\}$, describing the fluctuations in the large scale voltage instability of complex power networks

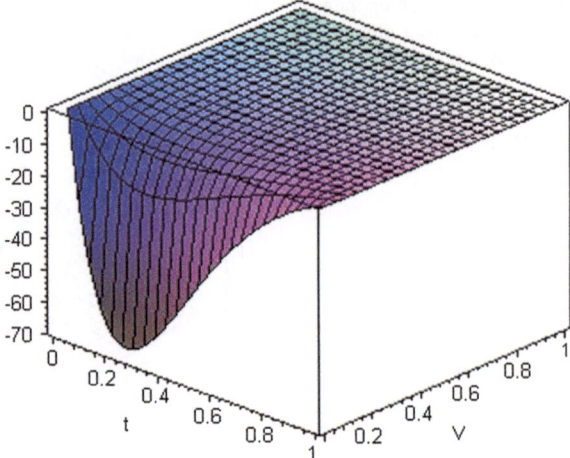

Fig. 9.4 The ab component of the metric tensor plotted as a function of $\{t, V\}$, describing the fluctuations in the large scale voltage instability of complex power networks

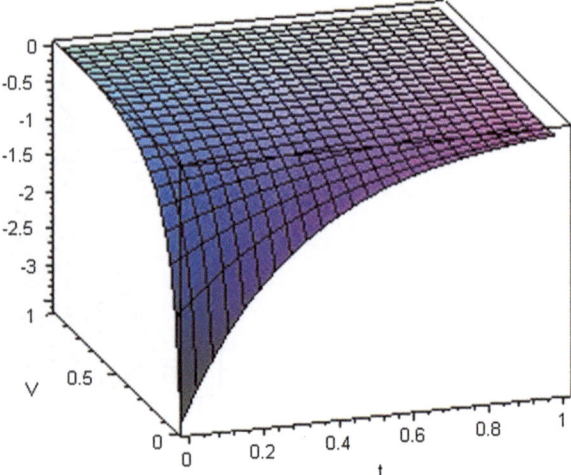

Similarly, when all the state-space parameters are allowed to fluctuate simultaneously, we find that stabilization requires the determinant of the metric tensor, viz.,

$$g(a, b, V) = \frac{g_n}{g_d}, \tag{9.6}$$

to be a positive quantity on the fluctuation manifold (M_3, g). Here, the factors $\{g_n, g_d\}$ of the determinant of the metric tensor g can be expressed in the form

Fig. 9.5 The aV component
of the metric tensor plotted as
a function of $\{t, V\}$, describing
the fluctuations in the large
scale voltage instability of
complex power networks

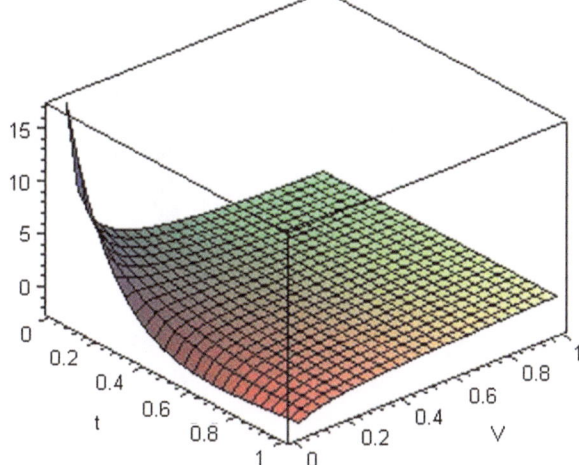

Fig. 9.6 The bV component
of the metric tensor plotted as
a function of $\{t, V\}$, describing
the fluctuations in the large
scale voltage instability of
complex power networks

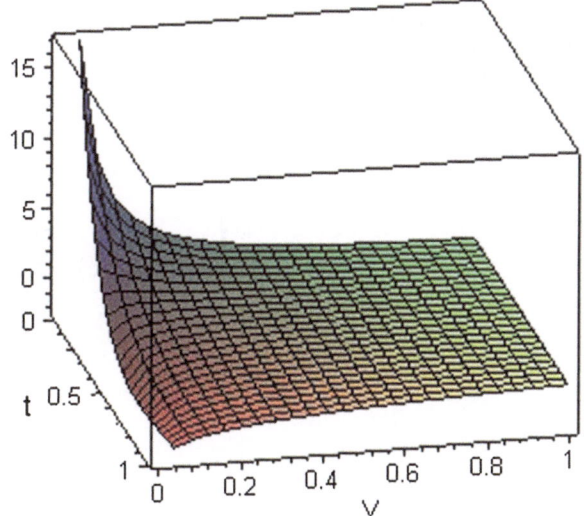

$$
g_n = -2Q_0^2 \ln\left(\frac{V}{V_0}\right)^2 P_0^2 \left[\left(\frac{V}{V_0}\right)^{2a+4b} Q_0^2 + 2\left(\frac{V}{V_0}\right)^{2a+4b} Q_0^2 \ln\left(\frac{V}{V_0}\right) b \right.
$$
$$
+ \left(\frac{V}{V_0}\right)^{2a+4b} Q_0^2 \ln\left(\frac{V}{V_0}\right)^2 b + \left(\frac{V}{V_0}\right)^{4a+2b} \ln\left(\frac{V}{V_0}\right)^2 P_0^2 a
$$
$$
\left. + 2\left(\frac{V}{V_0}\right)^{4a+2b} P_0^2 \ln\left(\frac{V}{V_0}\right) a + \left(\frac{V}{V_0}\right)^{4a+2b} P_0^2 \right],
$$
$$
g_d = V^2 \left[P_0^2 \left(\frac{V}{V_0}\right)^{2a} + Q_0^2 \left(\frac{V}{V_0}\right)^{2b} \right]^{3/2} .
$$

(9.7)

Fig. 9.7 Determinant of the
metric tensor plotted as a
function of $\{t, V\}$, describing
the fluctuations in the large
scale voltage instability of
complex power networks

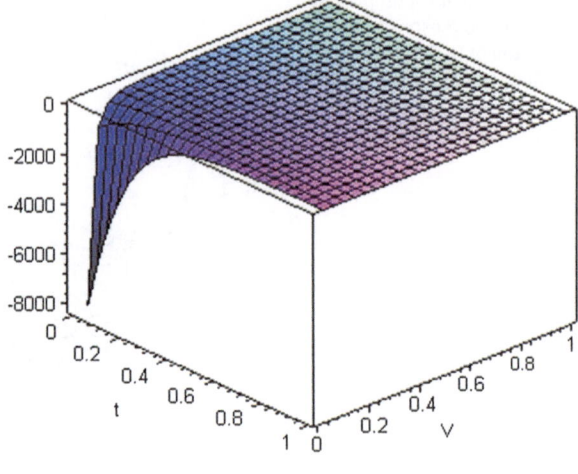

Fig. 9.8 Surface minor of
the metric tensor plotted as a
function of $\{t, V\}$, describing
the fluctuations in the large
scale voltage instability of
complex power networks

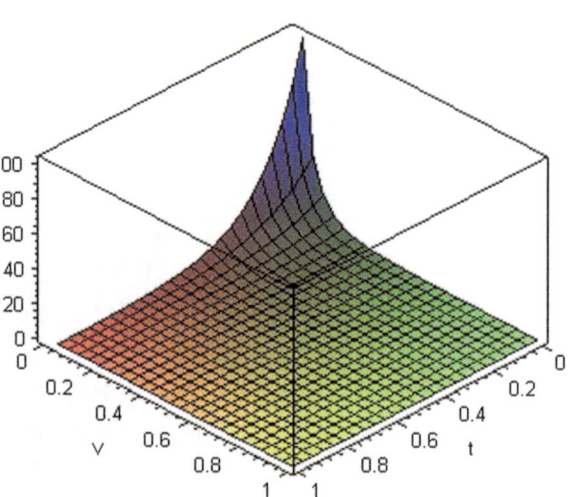

As a function of $\{t, V\}$, the ensemble stability condition of large voltage instability
networks follows from the positivity of the determinant of the metric tensor. In this
case, we find that the determinant of the metric tensor generically tends to a negative
value. For a typical value of $t \in (0, 1)$ and $V \in (0, 1)$, Fig. 9.7 shows that the
determinant of the metric tensor lies in the interval $(-8000, 0)$. For small t, we
observe that the determinant of the metric tensor has an approximate value of -8000
as we decrease the value of the output voltage V. In the limit of large t, the determinant
of the metric tensor remains almost constant for a large value of V. Hence, in the limit
of small t and small S, it increases sharply to a large negative value of order -8000.
The corresponding stability of the surface defined by a constant value of the voltage
is shown in Fig. 9.8. In this range of $\{t, V\}$, we see that the minor p_2 lies in the range

Fig. 9.9 Scalar curvature
plotted as a function of $\{t, V\}$,
describing the fluctuations
in the large scale voltage
instability of complex power
networks

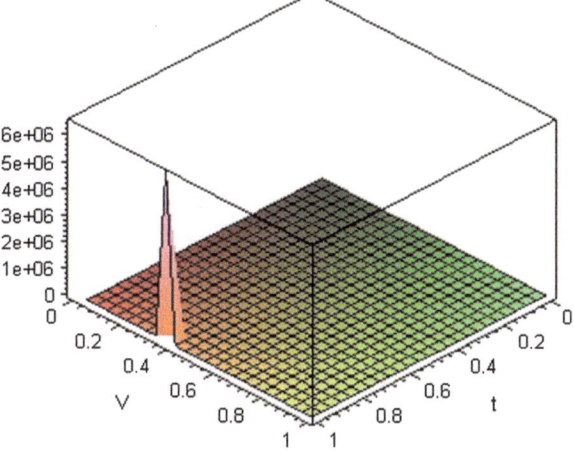

(0, 100). In the limit of small t, we find that the positivity of p_2 increases as the value
of V is decreased from 1 to zero. When a is the only parameter which is allowed
to vary, the stability of the configuration is determined by the positivity of the first
principal minor $p_1 := g_{aa}$. Thus, the above graphical descriptions of the principal
minors provide a qualitative notion of the stability of large scale voltage-unstable
networks.

Note further that we have an exact formula for the global scalar curvature of
the fluctuation. In particular, it can be shown directly that we have the following
expression for the scalar curvature:

$$R(a, b, V) = \frac{1}{2}\sqrt{P_0^2 \left(\frac{V}{V_0}\right)^{2a} + Q_0^2 \left(\frac{V}{V_0}\right)^{2b}} \, \frac{R_n^{(1)} + R_n^{(2)} + R_n^{(3)}}{R_d^{(1)} + R_d^{(2)}}, \qquad (9.8)$$

where the factors $\{R_n^{(1)}, R_n^{(2)}, R_n^{(3)}\}$ in the numerator and $\{R_d^{(1)}, R_d^{(2)}\}$ in the denominator of the scalar curvature R take a set of simple forms. In order to maintain the
readability of the presentation, we relegate the complete expression of this scalar
curvature to Appendix D.

Under fluctuations of the parameters $\{a, b\}$ and output voltage V, the global
stability properties of large instability voltage networks are shown in Fig. 9.9. In the
range $t \in (0, 1)$ and $V \in (0, 1)$, it follows from Fig. 9.9 that the Ricci scalar curvature
reaches a large positive peak. In fact, in the limit of large t, Fig. 9.9 shows that the
scalar curvature has certain highly interacting domains around $V = 0.4$. In particular,
in this range of the flow parameters, we see that the scalar curvature has a positive
amplitude of the order $6 \times 10^{+6}$. Thus, we may note that the underlying network
becomes a strongly interacting system in this limit of equal flow index and output
voltage. As mentioned in the case of complex power optimization, fluctuations in
the flow index and output voltage yield a positive sign for the scalar curvature. This

signifies that large voltage instability power law networks are repulsive in nature. Qualitatively, the above graphical depictions of the scalar curvatures indicate that large voltage instability networks are unstable configurations only in a small domain of the parameters. As expected, we find that power law modeling corresponds to well-behaved, weakly interacting, and relatively stable network configurations.

Chapter 10
Conclusion and Outlook

The present research concerns the problem of planning in the power industry. The issues of network reliability and bus voltage stability have been examined from the perspective of engineering applications to planning and operation using intrinsic geometric considerations. Specifically, we have considered an intrinsic geometric model for the limiting reliability and limiting voltage stability analysis of electrical networks. The correlation model has been converted into an intrinsic geometric model in order to cater for the non-linear effects of stochastic nature encoded in power systems with an appropriate optimization of the components. The robustness of the proposed model is illustrated by introducing variations of the circuit parameters. In this model, reliability and stability of a component are directly accomplished through the framework of intrinsic Riemannian geometry.

Starting from the definition of the present approach, we have presented a systematic exposition and conversion of the basic principles of covariance and correlation techniques bringing together: (i) the intrinsic geometric model, (ii) advances in power network analysis, (iii) system stability, and (iv) high efficiency and performance. These considerations have proven the viability of our proposal for dealing with non-linear effects. It is worth mentioning that the speed of the proposed analysis for choosing compensation strategies, transients, and power flows subject to variations of the resistive and inductive loads can be made arbitrarily high. The power flows include the consequences arising from the real and imaginary power of an electrical circuit with finitely many nodes. The notion of criticality is explicated by determining critical exponents of the global reliability and voltage stability. This investigation can establish the network reliability and voltage stability of the buses under local variations of the load impedance(s). Such a methodology provides a way for the power industry to re-engineer network designs with improved insurance for the power system.

In a nutshell, the proposed intrinsic geometric model has the following advantages:

- It provides a global picture of the network reliability and busbar voltage stability.
- It can deal with non-linear effects of the network reliability and busbar voltage stability.
- It can handle transient phenomena.

S. Bellucci et al., *Geometrical Methods for Power Network Analysis*,
SpringerBriefs in Electrical and Computer Engineering,
DOI: 10.1007/978-3-642-33344-6_10, © The Author(s) 2013

This model further supports detailed analysis and studies pertaining to the prevention of large-scale blackouts. With such considerations, we can efficiently maintain the economic viability of the underlying configuration and thus guarantee an accessible power market. Furthermore, in accordance with existing network planning strategies, our geometric proposal may provide suitable locations for constructing generation plants. One can further speed up the operating capability for network owners in order to keep the desired network efficient while improving the performance of the power system. The proposed methodology should be able to accompany future research into optimal electricity market designs and planning.

We have studied power system planning in which voltage stability is the main concern. In this setup, advancements thus introduced into the electricity market involve radical changes to power system structures. In the electricity sector of the power industry, we have considered an intrinsic geometric model to make a power system non-linearly efficient. Our model gives promising optimization criteria to select the optimal network parameters. The robustness of the model has been illustrated by variation(s) of the impedance angles, viz., phases. In terms of impedance, resistance, and capacitance, our construction describes a definite stability character with respect to network power fluctuations. As a function of certain trigonometric polynomials, we have demonstrated that canonical fluctuations can be accurately depicted without approximation.

In this book, we have analyzed the statistical fluctuations in the real and imaginary power flows and thus characterized the network configurations. Furthermore, for the joint effect of the real and imaginary power equations, we have shown that the intrinsic geometric notion offers a clear picture of fluctuating network parameters. Such a configuration, as the limit of an ensemble of power factor fluctuations, reduces to a specific electrical network. The present analysis does not stop here, but investigates the nature of the underlying imaginary power flow and explicates the global stability for identical power factors. Our study thus offers an appropriate network design, ensuring the stability of existing linear optimization techniques.

Finally, our proposal provides suitable tests for planning the network parameters of finite component power systems. In our analysis, the intrinsic geometric model takes into account the non-linear effects arising in stochastic power systems. A novel approach is thereby made possible in the field of network design. The criteria deduced from our method can be used to determine optimization constraints in the economic analysis of power systems, but also for setting up the operating point(s) of the power system, viz., network planning and compensation techniques. This task is left for future investigation.

The intrinsic geometric analysis follows the state-space approach to investigating power system instability, showing that this modeling approach is very useful for fine-tuning the operating point, because it incorporates fluctuations in the decision variables. In this research, three issues are demonstrated and analyzed for the case of the two-bus power system. The use of the Hessian matrix in the proposed model improves the ability to predict the real aspects of electrical power systems after faults and contingencies, as compared to existing models based on linear Jacobian matrix analysis. We note that the configurations under consideration are allowed to be effectively attractive or repulsive, and weakly interacting. In general, the intrinsic geometric analysis further provides a set of physical indications encoded in the geometrically invariant scalar curvature. For electrical networks, the underlying analysis would involve an ensemble of the equilibrium configuration, forming a statistical basis about the Gaussian distribution.

The whole process can be applied to a complex power network with many transmission lines and buses. The solution of the regulation problem by the present methodology shows that phase shift correction actions can be improved. The precise solution can deal with the instability issue by taking decisions in positive determinant regions of the network parameters. In a practical power system, the power flow occurs at some dropping voltage, and the present proposal ensures the limiting instability. For a given generator voltage, the analysis of fluctuations is well suited to determining the allowed operating regimes. As regards the voltage regulation problem, we showed in Chap. 7 how to determine the phase correction s, for a specific set of fluctuations in t, corresponding to metric tensor determinant values exceeding the limits of the given complex power. In order to ensure the stability of the power system, the present analysis is carried out by taking the complex power instead of only the real power.

Voltage instability studies lead to complex power optimization and deliver voltage stable solutions for the power system. The required real and reactive power is selected from the map based on differential geometry [see, e.g., (2.13)]. This map represents the power in terms of the fluctuating network parameters. Further, the long term voltage stability issue is solved by using a load model. The value of the uniform index parameter t represents the effect of different types of load model, corresponding to different exponents. Regarding the behavior of the power system for different values of a and b and the associated voltage to keep the power system stable, and in particular for a given complex power, we have provided a non-linear solution to the network optimization theory. Thus, the described intrinsic geometric modeling can improve the operation of the power system under the highly deregulated conditions of power market operation. The analysis shows the operation of the power system in both domains defined by $k = \pm 1$, for both low voltage and high voltage profiles of the given network parameters. Sensitivity analysis of the network parameters allows planners to select the optimal value of the transmission line parameters, e.g., the capacity and the length of the lines.

This approach can be further utilized in the planning process, where the siting of a new power plant can be analyzed in terms of the power system instability. The detailed computations and plots of the underlying state-space quantities offer a modern cutting-edge approach for examining and controlling the performance of the deregulated power market. In order to realize the above scenarios, the present methodology depends on the nature of the model parameters. By considering a given set of (input) parameters, we propose a precise and detailed example of network planning. We extend the state-space analysis to non-steady state regimes and large scale voltage instability. Moreover, the addition of a set of new transmission lines can be analyzed in terms of their capacities and lengths. The additional bus shunt capacitance can be provided with the help of the present approach. The generalized application of the above method could be explored further for an m-bus system with n-transmission lines. Note that it would be possible to select the new transmission lines from the given alternatives.

When designing power systems, and in particular siting nuclear power plants and renewable energy sources, a stability and feasibility study is required, and this model would be useful to analyze the stability of the network for the planned power generation. This method can provide an accurate load shedding strategy with respect to system instability.

Appendix A
Factors of the Scalar Curvature for Complex Power Optimization

In this appendix, we provide the specific factors of the scalar curvature. In particular, we find after simplification that the factors r_{10}, r_{11}, r_{12}, r_{13}, r_{14}, and r_{15} are as follows:

$$r_{10} = 30720y^4zx^6 - 8192zy^6x^6 - 49152x^7zy^2 + 6768zy^2x^3 + 2304x^5zy^2, \quad (A.1)$$

$$\begin{aligned}
r_{11} = {} & 259200zy^2x^2 + 66240zy^6 - 172800x^4z + 122112x^5z \\
& + 737280zy^6x^2 - 26640zx^2 - 610560zy^2x^3 + 41472y^{10}z \\
& - 92160y^8z + 3645zx + 96480zx^3 - 198z + 898560zx^3y^4 \\
& - 449280zy^6x + 3330zy^2 - 794880zy^4x^2 + 288000y^8zx \\
& + 529920x^4zy^2 - 48240zy^2x - 21600zy^4 + 228960zy^4x, \quad (A.2)
\end{aligned}$$

$$\begin{aligned}
r_{12} = {} & -1134z - 4757744zy^2x^3 - 4323072x^5zy^2 - 114688y^{10}zx \\
& - 1373184y^8zx^2 + 1030400y^8zx + 43008y^{10}z - 197504y^8z \\
& + 8192y^{12}z + 1557672zy^2x^2 - 1714368zy^6x + 4791168zy^6x^2 \\
& + 7198208x^4zy^2 - 252212zy^2x - 6422272x^4zy^4 - 4528992zy^4x^2 \\
& + 8832576zx^3y^4 + 1022992zy^4x - 4463872zy^6x^3 - 85608zy^4 \\
& + 203296zy^6 + 742380zx^3 + 21990zx - 175854zx^2 + 2169408x^5z \\
& - 1745616x^4z - 1114368x^6z + 16142zy^2, \quad (A.3)
\end{aligned}$$

S. Bellucci et al., *Geometrical Methods for Power Network Analysis*,
SpringerBriefs in Electrical and Computer Engineering,
DOI: 10.1007/978-3-642-33344-6, © The Author(s) 2013

$$
\begin{aligned}
r_{13} = {} & -882z - 4590576zy^2x^3 - 11378688x^5zy^2 + 163840y^{10}zx \\
& + 104448y^8zx^2 - 122880y^{10}zx^2 + 163840y^8zx^3 - 143104y^8zx \\
& - 47616y^{10}z + 29440y^8z + 1200984zy^2x^2 + 2793984zy^6x^4 \\
& - 414912zy^6x + 5467648y^2zx^6 + 1820800zy^6x^2 + 6289920x^5zy^4 \\
& + 9886304x^4zy^2 - 167684zy^2x - 10947200x^4zy^4 - 2721696zy^4x^2 \\
& + 7681344zx^3y^4 + 487472zy^4x - 3659008zy^6x^3 - 35280zy^4 \\
& + 37248zy^6 + 798460zx^3 + 18336zx - 162622zx^2 + 4128736x^5z \\
& + 1684992x^7z - 2346168x^4z - 4031104x^6z + 9746zy^2, \qquad\qquad \text{(A.4)}
\end{aligned}
$$

$$
\begin{aligned}
r_{14} = {} & -90z - 841728y^4zx^6 - 1245184x^7zy^2 - 484496zy^2x^3 \\
& - 2784768x^5zy^2 + 122880y^8zx^4 + 285696y^8zx^2 - 327680y^8zx^3 \\
& - 102144y^8zx + 12928y^8z + 94296zy^2x^2 - 473088zy^6x^4 \\
& + 16384zy^6x^5 + 103104zy^6x + 2856960y^2zx^6 - 409472zy^6x^2 \\
& + 1328640x^5zy^4 + 1500160x^4zy^2 - 10300zy^2x - 728064x^4zy^4 \\
& + 25440zy^4x^2 + 123072zx^3y^4 - 11376zy^4x + 716032zy^6x^3 \\
& + 1080zy^4 - 9632zy^6 + 127268zx^3 + 2106zx - 21626zx^2 \\
& + 1109152x^5z + 1383936x^7z - 509952x^8z - 469248x^4z \\
& - 1639552x^6z + 490zy^2, \qquad\qquad\qquad\qquad\qquad\qquad\quad \text{(A.5)}
\end{aligned}
$$

$$
\begin{aligned}
r_{15} = {} & -1752zy^2x^2 - 47616zy^6x^4 + 32768zy^6x^5 + 2496zy^6x \\
& + 30720y^2zx^6 - 12928zy^6x^2 - 73728x^5zy^4 - 12384x^4zy^2 \\
& + 24576x^8zy^2 + 228zy^2x + 72576x^4zy^4 + 10656zy^4x^2 \\
& - 37440zx^3y^4 - 1584zy^4x + 34048zy^6x^3 + 96zy^4 - 192zy^6 \\
& + 788zx^3 + 3zx - 74zx^2 + 17472x^5z + 58368x^7z - 47104x^8z \\
& - 4728x^4z + 16384x^9z - 40704x^6z - 12zy^2. \qquad\qquad\qquad \text{(A.6)}
\end{aligned}
$$

Appendix B
Tilde Factors of the Scalar Curvature for Complex Power Optimization

In this appendix, we provide the specific tilde factors of the scalar curvature, viz., $\tilde{r}_{11}$, $\tilde{r}_{12}$, $\tilde{r}_{13}$, $\tilde{r}_{14}$, and $\tilde{r}_{15}$:

$$
\begin{aligned}
\tilde{r}_{11} = {} & 138240y^2x^4 + 51840y^4x - 4320x^2 - 138240y^6x - 4320y^4 \\
& + 17280y^6 + 51840y^2x^2 + 540y^2 - 207360y^4x^2 + 17280x^3 \\
& + 27648x^5 + 138240y^8x + 540x - 138240y^2x^3 - 27 - 8640y^2x \\
& - 34560y^8 - 34560x^4 + 276480y^6x^2 + 27648y^{10} + 276480y^4x^3, \quad \text{(B.1)}
\end{aligned}
$$

$$
\begin{aligned}
\tilde{r}_{12} = {} & -630 + 11070y^2 + 13146x - 3691008y^8x^2 - 4918464y^4x^2 \\
& - 188232y^2x - 804864y^{10}x - 4460544y^2x^5 - 7698432y^6x^3 \\
& + 1009344y^4x - 4203648y^2x^3 + 6890496y^2x^4 - 2455296y^6x \\
& + 7595520y^6x^2 + 1265760y^2x^2 + 2554368y^8x - 8260608y^4x^4 \\
& + 10486272y^4x^3 - 76512y^4 - 1307328x^4 + 1747968x^5 - 963072x^6 \\
& + 515808x^3 + 259392y^6 - 433152y^8 - 49152y^{12} + 302592y^{10} \\
& - 113298x^2, \quad \text{(B.2)}
\end{aligned}
$$

$$
\begin{aligned}
\tilde{r}_{13} = {} & -1260 + 18598y^2 + 27738x + 13643264x^6y^2 - 5578240y^8x^2 \\
& - 8801920y^4x^2 + 5292032y^8x^3 - 340104y^2x - 133120y^{10}x \\
& - 27554560y^2x^5 + 360448y^{10}x^2 - 22633472y^6x^3 + 1510432y^4x \\
& + 21540352x^5y^4 - 10308288y^2x^3 + 23121888y^2x^4 + 16071168y^6x^4 \\
& - 98304y^{12}x - 2870784y^6x + 12049664y^6x^2 + 2572048y^2x^2 \\
& + 2021376y^8x - 37100416y^4x^4 + 25568768y^4x^3 - 103112y^4 \\
& - 4131240x^4 + 7589600x^5 - 7708032x^6 + 1341600x^3 + 257312y^6 \\
& - 253312y^8 + 65536y^{12} + 8192y^{10} + 3340800x^7 - 259730x^2, \quad \text{(B.3)}
\end{aligned}
$$

S. Bellucci et al., *Geometrical Methods for Power Network Analysis*,   91
SpringerBriefs in Electrical and Computer Engineering,
DOI: 10.1007/978-3-642-33344-6, © The Author(s) 2013

$$
\begin{aligned}
\tilde{r}_{14} = {}& -378 + 4106y^2 + 9006x + 18238464x^6y^2 + 1376768y^8x^2 \\
& + 81920y^8x^4 - 9121792x^6y^4 - 1643456y^4x^2 - 3946496y^6x^5 \\
& - 1265664y^8x^3 - 83448y^2x - 7714816x^7y^2 + 380928y^{10}x \\
& - 18501760y^2x^5 - 655360y^{10}x^2 - 2601984y^6x^3 + 223968y^4x \\
& + 17999360x^5y^4 - 3557696y^2x^3 + 10454848y^2x^4 + 327680y^{10}x^3 \\
& + 5336064y^6x^4 - 34176y^6x + 530432y^6x^2 + 729472y^2x^2 \\
& - 506880y^8x - 14810880y^4x^4 + 6540032y^4x^3 - 12976y^4 \\
& - 2067792x^4 + 4912288x^5 - 7292928x^6 + 556992x^3 - 768y^6 \\
& + 61696y^8 - 68096y^{10} + 6185472x^7 - 2293760x^8 - 93734x^2, \quad \text{(B.4)}
\end{aligned}
$$

$$
\begin{aligned}
\tilde{r}_{15} = {}& -9 - 10y^2 + 258x + 1674752x^6y^2 + 204288y^8x^2 + 327680y^8x^4 \\
& - 581632x^6y^4 + 86016y^4x^2 + 671744x^8y^2 + 587776y^6x^5 \\
& - 380928y^8x^3 - 120y^2x - 1636352x^7y^2 - 930944y^2x^5 \\
& + 747520y^6x^3 - 13376y^4x + 21504x^5y^4 - 56192y^2x^3 \\
& + 301792y^2x^4 - 1009152y^6x^4 + 51840y^6x - 65536y^6x^6 \\
& - 279296y^6x^2 + 5264y^2x^2 - 51712y^8x + 367232y^4x^4 \\
& - 98304y^8x^5 + 434176x^7y^4 - 268544y^4x^3 + 824y^4 \\
& - 116040x^4 + 366720x^5 - 768384x^6 + 24384x^3 - 3808y^6 \\
& + 4992y^8 + 1027584x^7 - 794624x^8 + 270336x^9 - 3286x^2. \quad \text{(B.5)}
\end{aligned}
$$

Appendix C
Components of the Metric Tensor for Large Scale Voltage Instability

In this appendix, we give explicit expressions for the fluctuation components. From the definition of the state-space geometry, we find for parameters $\{a, b, V\}$ that the components of the metric tensor can be expressed by the following two sets of equations. The components involving the parameters $\{a, b\}$ are:

$$g_{aa} = \frac{P_0^2 \ln\left(\frac{V}{V_0}\right)^2 \left[\left(\frac{V}{V_0}\right)^{4a} P_0^2 + 2\left(\frac{V}{V_0}\right)^{2a+2b} Q_0^2\right]}{\left[P_0^2 \left(\frac{V}{V_0}\right)^{2a} + Q_0^2 \left(\frac{V}{V_0}\right)^{2b}\right]^{3/2}}, \qquad (C.1)$$

$$g_{ab} = -\frac{P_0^2 \left(\frac{V}{V_0}\right)^{2a+2b} \ln\left(\frac{V}{V_0}\right)^2 Q_0^2}{\left[P_0^2 \left(\frac{V}{V_0}\right)^{2a} + Q_0^2 \left(\frac{V}{V_0}\right)^{2b}\right]^{3/2}}, \qquad (C.2)$$

$$g_{bb} = \frac{Q_0^2 \ln\left(\frac{V}{V_0}\right)^2 \left[\left(\frac{V}{V_0}\right)^{4b} Q_0^2 + 2\left(\frac{V}{V_0}\right)^{2a+2b} P_0^2\right]}{\left[P_0^2 \left(\frac{V}{V_0}\right)^{2a} + Q_0^2 \left(\frac{V}{V_0}\right)^{2b}\right]^{3/2}}. \qquad (C.3)$$

Those involving the parameter V can be written as follows. The aV component is

$$g_{aV} = \frac{g_{aV}^{(n)}}{g_{aV}^{(d)}}, \qquad (C.4)$$

where the factors $g_{aV}^{(n)}$ and $g_{aV}^{(d)}$ are given by

S. Bellucci et al., *Geometrical Methods for Power Network Analysis*,
SpringerBriefs in Electrical and Computer Engineering,
DOI: 10.1007/978-3-642-33344-6, © The Author(s) 2013

$$
g_{aV}^{(n)} = P_0^2 \left[\left(\frac{V}{V_0} \right)^{4a} a \ln \left(\frac{V}{V_0} \right) P_0^2 - \left(\frac{V}{V_0} \right)^{2a+2b} \ln \left(\frac{V}{V_0} \right) Q_0^2 b \right. \tag{C.5}
$$
$$
\left. + 2 \left(\frac{V}{V_0} \right)^{2a+2b} a \ln \left(\frac{V}{V_0} \right) Q_0^2 + \left(\frac{V}{V_0} \right)^{4a} P_0^2 + \left(\frac{V}{V_0} \right)^{2a+2b} Q_0^2 \right],
$$

$$
g_{aV}^{(d)} = \left[P_0^2 \left(\frac{V}{V_0} \right)^{2a} + Q_0^2 \left(\frac{V}{V_0} \right)^{2b} \right]^{3/2} V. \tag{C.6}
$$

Likewise, the bV component takes the form

$$
g_{bV} = \frac{g_{bV}^{(n)}}{g_{bV}^{(d)}}, \tag{C.7}
$$

where $g_{bV}^{(n)}$ and $g_{bV}^{(d)}$ are given by

$$
g_{bV}^{(n)} = -Q_0^2 \left[\left(\frac{V}{V_0} \right)^{2a+2b} a \ln \left(\frac{V}{V_0} \right) P_0^2 - \left(\frac{V}{V_0} \right)^{4b} \ln \left(\frac{V}{V_0} \right) Q_0^2 b \right. \tag{C.8}
$$
$$
\left. - 2 \left(\frac{V}{V_0} \right)^{2a+2b} b \ln \left(\frac{V}{V_0} \right) P_0^2 - \left(\frac{V}{V_0} \right)^{2a+2b} P_0^2 - \left(\frac{V}{V_0} \right)^{4b} Q_0^2 \right],
$$

$$
g_{bV}^{(d)} = \left[P_0^2 \left(\frac{V}{V_0} \right)^{2a} + Q_0^2 \left(\frac{V}{V_0} \right)^{2b} \right]^{3/2} V. \tag{C.9}
$$

Finally, the VV component of the metric tensor is given by

$$
g_{VV} = \frac{g_{VV}^{(n)}}{g_{VV}^{(d)}}, \tag{C.10}
$$

where the factors $g_{VV}^{(n)}$ and $g_{VV}^{(d)}$ are

$$
g_{VV}^{(n)} = P_0^4 \left(\frac{V}{V_0} \right)^{4a} a^2 - 2P_0^2 \left(\frac{V}{V_0} \right)^{2a+2b} a Q_0^2 b + Q_0^4 \left(\frac{V}{V_0} \right)^{4b} b^2
$$
$$
+ 2P_0^2 \left(\frac{V}{V_0} \right)^{2a+2b} a^2 Q_0^2 - P_0^4 \left(\frac{V}{V_0} \right)^{4a} a - P_0^2 \left(\frac{V}{V_0} \right)^{2a+2b} a Q_0^2
$$
$$
+ 2Q_0^2 \left(\frac{V}{V_0} \right)^{2a+2b} b^2 P_0^2 - Q_0^2 \left(\frac{V}{V_0} \right)^{2a+2b} b P_0^2 - Q_0^4 \left(\frac{V}{V_0} \right)^{4b} b \quad \text{(C.11)}
$$

and

$$
g_{VV}^{(d)} = \left[P_0^2 \left(\frac{V}{V_0} \right)^{2a} + Q_0^2 \left(\frac{V}{V_0} \right)^{2b} \right]^{3/2} V^2. \tag{C.12}
$$

Appendix D
Scalar Curvature for Large Scale Voltage Instability

In this appendix, we give an exact formula for the characterization of global fluctuation properties. As in (9.8) on p. xxx, we find that the factors $R_n^{(1)}$, $R_n^{(2)}$, and $R_n^{(3)}$ in the numerator of the scalar curvature R are given by

$$
\begin{aligned}
R_n^{(1)} = {} & 2Q_0^2 \left(\frac{V}{V_0}\right)^{2a+2b} \ln\left(\frac{V}{V_0}\right)^2 P_0^2 - 4\ln\left(\frac{V}{V_0}\right)^3 Q_0^2 \left(\frac{V}{V_0}\right)^{2a+2b} P_0^2 a \\
& - 4Q_0^2 \left(\frac{V}{V_0}\right)^{2a+2b} \ln\left(\frac{V}{V_0}\right)^3 bP_0^2 - 15\ln\left(\frac{V}{V_0}\right)^2 Q_0^2 \left(\frac{V}{V_0}\right)^{2a+2b} P_0^2 a \\
& - 16Q_0^2 \left(\frac{V}{V_0}\right)^{2a+2b} P_0^2 - 15Q_0^2 \left(\frac{V}{V_0}\right)^{2a+2b} \ln\left(\frac{V}{V_0}\right)^2 bP_0^2 \\
& - 18Q_0^2 \left(\frac{V}{V_0}\right)^{2a+2b} P_0^2 \ln\left(\frac{V}{V_0}\right) a - 18Q_0^2 \left(\frac{V}{V_0}\right)^{2a+2b} \ln\left(\frac{V}{V_0}\right) bP_0^2 \\
& - 4Q_0^2 \left(\frac{V}{V_0}\right)^{2a+2b} P_0^2 \ln\left(\frac{V}{V_0}\right) - 8P_0^2 \left(\frac{V}{V_0}\right)^{2a+2b} \ln\left(\frac{V}{V_0}\right)^2 aQ_0^2 b \\
& - 8Q_0^2 \left(\frac{V}{V_0}\right)^{2a+2b} P_0^2 a \ln\left(\frac{V}{V_0}\right)^3 b - 2Q_0^2 \left(\frac{V}{V_0}\right)^{2a+2b} \ln\left(\frac{V}{V_0}\right)^4 bP_0^2 a \\
& - 2Q_0^4 \left(\frac{V}{V_0}\right)^{4b} \ln\left(\frac{V}{V_0}\right),
\end{aligned}
\tag{D.1}
$$

S. Bellucci et al., *Geometrical Methods for Power Network Analysis*,
SpringerBriefs in Electrical and Computer Engineering,
DOI: 10.1007/978-3-642-33344-6, © The Author(s) 2013

$$
\begin{aligned}
R_n^{(2)} =\ & \ln\left(\frac{V}{V_0}\right)^2 P_0{}^4 \left(\frac{V}{V_0}\right)^{4a} - 2P_0{}^4 \left(\frac{V}{V_0}\right)^{4a} \ln\left(\frac{V}{V_0}\right) + \ln\left(\frac{V}{V_0}\right)^2 Q_0{}^4 \left(\frac{V}{V_0}\right)^{4b} \\
& - 8Q_0{}^4 a \left(\frac{V}{V_0}\right)^{4b} b \ln\left(\frac{V}{V_0}\right)^3 - 8a\, P_0{}^4 b \left(\frac{V}{V_0}\right)^{4a} \ln\left(\frac{V}{V_0}\right)^3 \\
& - 8a \ln\left(\frac{V}{V_0}\right)^2 Q_0{}^4 \left(\frac{V}{V_0}\right)^{4b} b - 8P_0{}^4 \left(\frac{V}{V_0}\right)^{4a} \ln\left(\frac{V}{V_0}\right)^2 ab \\
& - 2Q_0{}^4 \left(\frac{V}{V_0}\right)^{4b} \ln\left(\frac{V}{V_0}\right)^4 ba - 2\ln\left(\frac{V}{V_0}\right)^4 P_0{}^4 \left(\frac{V}{V_0}\right)^{4a} ab - 8Q_0{}^4 \left(\frac{V}{V_0}\right)^{4b} \\
& - 8P_0{}^4 \left(\frac{V}{V_0}\right)^{4a} - 4\ln\left(\frac{V}{V_0}\right)^3 P_0{}^4 \left(\frac{V}{V_0}\right)^{4a} a - 13\ln\left(\frac{V}{V_0}\right)^2 Q_0{}^4 \left(\frac{V}{V_0}\right)^{4b} b \\
& + 4P_0{}^4 \left(\frac{V}{V_0}\right)^{4a} \ln\left(\frac{V}{V_0}\right)^2 a^2 - 14 Q_0{}^4 \left(\frac{V}{V_0}\right)^{4b} \left(\frac{V}{V_0}\right) b,
\end{aligned}
\tag{D.2}
$$

$$
\begin{aligned}
R_n^{(3)} =\ & 4Q_0{}^4 \left(\frac{V}{V_0}\right)^{4b} \ln\left(\frac{V}{V_0}\right)^2 b^2 - 4a \ln\left(\frac{V}{V_0}\right) Q_0{}^4 \left(\frac{V}{V_0}\right)^{4b} \\
& - 13\ln\left(\frac{V}{V_0}\right)^2 P_0{}^4 \left(\frac{V}{V_0}\right)^a a - 14 P_0{}^4 \left(\frac{V}{V_0}\right)^{4a} \ln\left(\frac{V}{V_0}\right) a \\
& - 4\ln\left(\frac{V}{V_0}\right) b P_0{}^4 \left(\frac{V}{V_0}\right)^{4a} - 2\ln\left(\frac{V}{V_0}\right)^2 P_0{}^4 \left(\frac{V}{V_0}\right)^{4a} b \\
& + 4a^2 P_0{}^4 \left(\frac{V}{V_0}\right)^{4a} \ln\left(\frac{V}{V_0}\right)^3 - 2Q_0{}^4 \left(\frac{V}{V_0}\right)^{4b} \ln\left(\frac{V}{V_0}\right)^2 a \\
& + 4Q_0{}^4 \left(\frac{V}{V_0}\right)^{4b} b^2 \ln\left(\frac{V}{V_0}\right)^3 - 4\ln\left(\frac{V}{V_0}\right)^3 Q_0{}^4 \left(\frac{V}{V_0}\right)^{4b} b \\
& + Q_0{}^4 \left(\frac{V}{V_0}\right)^{4b} \ln\left(\frac{V}{V_0}\right)^4 b^2 + \ln\left(\frac{V}{V_0}\right)^4 P_0{}^4 \left(\frac{V}{V_0}\right)^{4a} a^2.
\end{aligned}
\tag{D.3}
$$

Finally, the factors $R_d^{(1)}$ and $R_d^{(2)}$ in the denominator of R are given by

$$
\begin{aligned}
R_d^{(1)} =\ & 2Q_0{}^2 \left(\frac{V}{V_0}\right)^{2b} \ln\left(\frac{V}{V_0}\right) b + Q_0{}^2 \left(\frac{V}{V_0}\right)^{2b} \ln\left(\frac{V}{V_0}\right)^2 b + Q_0{}^2 \left(\frac{V}{V_0}\right)^{2b} \\
& + 2P_0{}^2 \left(\frac{V}{V_0}\right)^{2a} \ln\left(\frac{V}{V_0}\right) a + P_0{}^2 \left(\frac{V}{V_0}\right)^{2a} + \ln\left(\frac{V}{V_0}\right)^2 P_0{}^2 \left(\frac{V}{V_0}\right)^{2a} a \\
& + 2Q_0{}^4 \left(\frac{V}{V_0}\right)^{4b} \ln\left(\frac{V}{V_0}\right) b + \ln\left(\frac{V}{V_0}\right)^2 Q_0{}^4 \left(\frac{V}{V_0}\right)^{4b} b + Q_0{}^4 \left(\frac{V}{V_0}\right)^{4b}
\end{aligned}
\tag{D.4}
$$

D Scalar Curvature for Large Scale Voltage Instability

and

$$
\begin{aligned}
R_{\mathrm{d}}^{(2)} = {}& 2Q_0^2 \left(\frac{V}{V_0}\right)^{2a+2b} P_0^2 \ln\left(\frac{V}{V_0}\right) a + Q_0^2 \left(\frac{V}{V_0}\right)^{2a+2b} \ln\left(\frac{V}{V_0}\right)^2 b P_0^2 \\
& + 2Q_0^2 \left(\frac{V}{V_0}\right)^{2a+2b} P_0^2 + 2Q_0^2 \left(\frac{V}{V_0}\right)^{2a+2b} \ln\left(\frac{V}{V_0}\right) b P_0^2 \\
& + \ln\left(\frac{V}{V_0}\right)^2 Q_0^2 \left(\frac{V}{V_0}\right)^{2a+2b} P_0^2 a + P_0^4 \left(\frac{V}{V_0}\right)^{4a} \\
& + \ln\left(\frac{V}{V_0}\right)^2 P_0^4 \left(\frac{V}{V_0}\right)^{4a} a + 2P_0^4 \left(\frac{V}{V_0}\right)^{4a} \ln\left(\frac{V}{V_0}\right) a.
\end{aligned}
\tag{D.5}
$$